TEACHING GUIDE
Logic of Fractions

June Mark

E. Paul Goldenberg

Mary Fries

Jane M. Kang

Tracy Cordner

transition
to algebra
*make algebra
make sense*

Unit **8**

Logic of
Fractions

NSF Research-based
National Science
Foundation-funded

EDC Learning
transforms
lives.

firsthand
HEINEMANN
DEDICATED TO TEACHERS™

*first*hand
An imprint of Heinemann
361 Hanover Street
Portsmouth, NH 03801-3912
www.heinemann.com

Offices and agents throughout the world

Education Development Center, Inc.
43 Foundry Avenue
Waltham, MA 02453-8313
www.edc.org

Co-Principal Investigators and Project Directors: E. Paul Goldenberg and June Mark

Development and Research Team: Tracy Cordner, Mary Fries, Mari Halladay, Jane M. Kang, and Josephine Louie

Contributors: Cindy Carter, Susan Creighton, Jeff Downin, Doreen Kilday, Deborah Spencer, and Yu Yan Xu

 This material is based on work supported by the National Science Foundation under Grant No. ESI-0917958. Opinions expressed are those of the authors and not necessarily those of the Foundation.

Transition to Algebra, Unit 8: Logic of Fractions Teaching Guide
ISBN-13: 978-0-325-05322-6

Transition to Algebra Teacher Resources
ISBN-13: 978-0-325-05790-3

Transition to Algebra, Unit 8 Student Worktexts, 10-pack
ISBN-13: 978-0-325-05310-3

Transition to Algebra Student Worktexts, 10 Sets of All 12 Units
ISBN-13: 978-0-325-05791-0

Printed in the United States of America on acid-free paper

18 17 16 15 14   RRD   1 2 3 4 5

Unit **8**

Logic of Fractions

CONTENTS

Logic of Fractions

Learning Goals

By the end of Unit 8, students should be able to:

- Recognize equivalent forms of rational numbers and expressions.

- Use the number line and area models to build logic about multiplication with fractions.

- Use area models to multiply and divide rational expressions and numbers.

- Understand equations as relationships that can be scaled using multiplication or division.

- Represent proportional relationships using equations.

- Identify information about the rate at which a quantity changes by examining the graph of a proportional relationship.

B uilding on tools and strategies developed in earlier units, Unit 8 focuses on basic operations with rational numbers and rational expressions. Number lines and area models are used to help students make sense of additive and multiplicative operations with fractions. Mobile puzzles are used to build students' intuition about proportional reasoning. In the last lesson, students see how graphs can be used to depict proportional relationships. Throughout the unit, students make connections between numerical quantities and algebraic expressions in ways that extend their understanding of the logic of fractions.

Fraction Logic over Fraction Rules

It is common for students to confuse techniques they have learned about fractions (like when to "cross-multiply" and when to "multiply across"). Using rules without understanding the underlying properties and logic of fractions can make it seem like fractions are impossible to understand or that fractions follow different rules from other numbers. Unit 8 focuses on making sense of fractions, using familiar tools like the number line and area models. Students are given an opportunity to understand what fractions are, why they behave the way they do, and how to work with rational numbers and expressions. The goal is not to unlearn or undermine rules that students are familiar with but to provide a logical background as to *why* they work. If you find students (or yourself) simply resorting to poorly understood rules, take a moment to consider and discuss the *logic* of the action. Is there a similar problem with simpler numbers that can help you think about this one? Why do the techniques work?

The Number Line and Fraction Logic

In Unit 3: *Micro-Geography of the Number Line,* students extended their knowledge of the number line to locate and find distances between fractions. As is true with integers, situating fractions on the number line helps students consider the relative magnitudes of the numbers and the distances between them. It also reinforces the idea that fractions live on the number line with integers and can be understood and treated similarly. The following is an example problem. In this problem, students relate fractions to integers ($\frac{5}{3}$ is a third of 5), recognize equivalent numbers ($\frac{5}{3}$ and $\frac{10}{6}$ are located at the same point), and have opportunities to compare relative magnitudes (just as 5 is 5 times 1, $\frac{5}{3}$ is 5 times $\frac{1}{3}$; similarly, just as $\frac{1}{6}$ is half of $\frac{1}{3}$, $\frac{10}{6}$ is half of $\frac{10}{3}$).

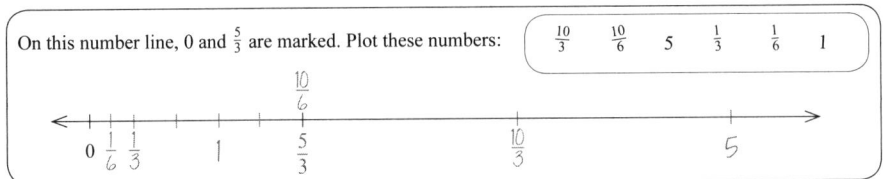

Area Models and Fraction Logic

Area models, first used in Unit 4: *Area and Multiplication,* help organize calculations for multiplication. Unit 8 helps students extend what they know to the multiplication of fractions with the aim of developing deeper understanding of both fractions and multiplication. The visual representation makes it clear, for example, how multiplying two fractions that are between 0 and 1 results in a product that is less than either factor. Students can also use the area model to show how the numerators and denominators of the factors affect the product. In doing so, students establish a logical understanding of why the product of fractions can be calculated by finding the product of the numerators and the product of the denominators.

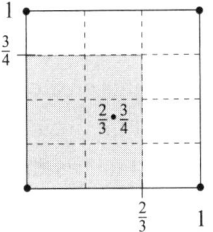

Scaling and Fraction Logic

As students learn that fraction multiplication is achieved by multiplying by the numerator and dividing by the denominator, they begin to see why dividing by 2 is the same as multiplying by $\frac{1}{2}$ and why multiplying a positive quantity by $\frac{6}{5}$ increases its value, while multiplying by $\frac{5}{6}$ decreases it. This reasoning will be extended to a variety of problems with fractions, including solving equations and adding fractions with unlike denominators.

Describing Repeated Reasoning

In this unit, students explore proportional relationships by first generating equivalent fractions and then using the same pattern of calculation to write an algebraic equation with variables. This general problem-solving strategy supports students in developing the perspective that algebra is a way of expressing general patterns found in numbers and the habit of using concrete cases to help them find those general patterns.

Mental Mathematics: Fractions, Approximation, and Factors

The Mental Mathematics activities in Unit 8 are organized around fractions and factoring. Students decide which fractions are less than, equal to, or greater than $\frac{1}{2}$ (and $\frac{1}{4}$), and they realize that counting the halves (or thirds) in a number is the same as multiplying by 2 (or 3). This thinking can change or deepen students' understanding of familiar fractions and builds the basis for making sense of division by fractions. Students are then guided to think about how many of *any* number can fit inside a larger one (e.g. how many 23's are in 1000). Such approximations are the first step in division. As students search mentally for factor pairs and common factors, they begin to develop the habit of looking at numbers as *products* of other numbers—complementing the more familiar decomposition of numbers as sums. These activities support student thinking about the relationship between fractions and division, multiplication and division as inverse operations, factoring, and equivalent fractions.

Explorations

The first Exploration, Triangular Numbers, culminates a series of Explorations in Units 1, 3, 6, and 7 that have involved this set of numbers in a variety of contexts. It is also part of the strand of Explorations about writing algebraic expressions for numerical patterns. Students now discover a formula for finding the nth triangular number by experimenting and describing their repeated reasoning.

Related Activity

Students create their own MysteryGrid puzzles to help them think more deeply about the structure, the nature of the clues, what makes a solution unique, and the strategies they can use to solve these puzzles—even those with harder clues such as fractions, which appear in this unit.

Lesson 1:
Rational Relationships

PURPOSE

This lesson lays foundations for the ways fractions are named and written and for proportional reasoning, especially using fractions. Lesson 1 lays the groundwork for all of Unit 8 by showing students that working with fractions is not about learning and memorizing rules, but about looking for and making sense of patterns. This initial work establishing logic with fractions by considering them numerical quantities will soon translate to work with fractions involving algebraic expressions.

Using Cuisenaire rods as a manipulative, a table of proportional relationships introduces the lesson. Students also express addition and subtraction with fractions on the number line.

Working with Cuisenaire rods highlights two main pieces of logic for students. First, the Cuisenaire rod activity supports students' understanding of proportional reasoning. If the white Cuisenaire rod is assigned a length of 5, then the purple rod, which is 4 white rods long, will have a length of 20. Likewise, if the white rod is assigned a length of one-third ($\frac{1}{3}$), then the purple rod will have a length of four-thirds ($\frac{4}{3}$), because it is 4 times as long. This understanding helps as students later make sense of finding $\frac{1}{3}b$ for various integer and fraction values of b. Second, students see a parallel between naming integers and naming fractions. Compared to rods that have lengths 1, 2, 3, 4, 5, and 6, the set of rods whose lengths are one-fourth of the originals will have lengths $\frac{1}{4}$, $\frac{2}{4}$, $\frac{3}{4}$, $\frac{4}{4}$, $\frac{5}{4}$, and $\frac{6}{4}$. This way of seeing fractions gives meaning to writing the fraction as "$\frac{5}{4}$" instead of changing it to $1\frac{1}{4}$ or some other form with equal value. This later helps students give meaning to expressions such as $\frac{n+1}{n}$.

 **Mental Mathematics** *Begin each day with five minutes of Mental Mathematics (pages T51–T64). Activities like "Comparing fractions to $\frac{1}{2}$" support a pattern-finding approach to exploring rational relationships.*

Launch: Cuisenaire Table and *Thinking Out Loud* Dialogue

Before students open their Worktexts, show or display a set of the first six Cuisenaire rods. Tell students the length of one of the rods, and ask them to use that information to figure out lengths for the other rods. Give students the following prompts:

- If the **red** block represents 8, find the values for the other rods.
  (The six rods will be: 4, 8, 12, 16, 20, and 24.)

- If the **purple** block represents 12, find the values for the other rods.
  (The six rods will be: 3, 6, 9, 12, 15, and 18.)

Preparation

· You will need at least one set of Cuisenaire rods, or you can use "Rod Manipulatives" (available in the Resources PDF and on page 55 of the Answer Key).

· *(optional)* Provide one set of Cuisenaire rods or cutouts for each group of students.

· Prepare to display the top of "Cuisenaire Tables" (available in the Resources PDF and on page T39) during the *Thinking Out Loud* dialogue.

Mental Mathematics (5 min)

Launch: Cuisenaire Table and *Thinking Out Loud* Dialogue (20 min)

· Describe and go through several rounds of identifying relative lengths using Cuisenaire rods.

· Provide time for students to begin filling out the table in problem 1.

· Have volunteers act out the dialogue.

· Discuss the different ways the characters in the dialogue make sense of the quantity $\frac{3}{2}$.

Student Problem Solving and Discussion (20 min)

· Allow students to work through the rest of the Important Stuff and explore additional problems.

· Discuss how equivalent fractions can be useful when adding and subtracting fractions.

Unit 8 Related Activity: Making MysteryGrids (See page T37 and Student Worktext page 50.)

Using physical Cuisenaire rods facilitates conversations 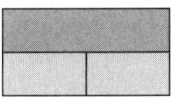 about relationships between the lengths of rods, such as showing that the bottom rod is exactly half the length of the top rod.

Students who seek and use structure look for relationships between numbers. With more than one way to think about a number, they can make sense of the quantities they encounter with greater flexibility.

The number 3, for example, has many identities: it is precisely halfway between 2 and 4; it is $3 \cdot 1$; it is $\frac{1}{2}$ of 6; it is $\frac{3}{2}$ of 2; it is $\frac{1}{4}$ of 12; it is $\frac{12}{4}$; and so on.

Likewise, the number $\frac{4}{3}$ has many identities: it is between 1 and 2 (precisely one-third of the way from 1 to 2); it is $4 \cdot \frac{1}{3}$; it is $1 + \frac{1}{3}$; it is $\frac{1}{3}$ of 4; it is $\frac{1}{2}$ of $\frac{8}{3}$; it is $\frac{1}{6}$ of 8; it is $4 \div 3$; and so on.

As students become more flexible in their thinking, they can look to the structure of a problem and use that structure to help them form a strategy for solving the problem. In this way, giving students the tools to help them make sense of quantities is more helpful than providing them with rules they can only apply rigidly.

Listen for students who compare the relative lengths of the Cuisenaire rods to justify their responses. For example, a student might say, "The dark green rod is three times the length of the red rod, so if red is 8, then dark green is 24," or "The red rod is half the length of the purple rod, so if purple is 12, then red is 6."

Some students may be lured, at first, to see the rods only as an ordered sequence so, because white, red, and light green are the first three in order, they will interpret "red is 8" to mean that white must be 7 and light green must be 9. Generally, though, even with such initial expectations, students readily accept that a better interpretation of the relationship of the rods is that "red is 2 whites" so if red is 8, then white must be 4 and light green must be 12.

Provide time for students to work on **PROBLEM 1** in the Student Worktext.

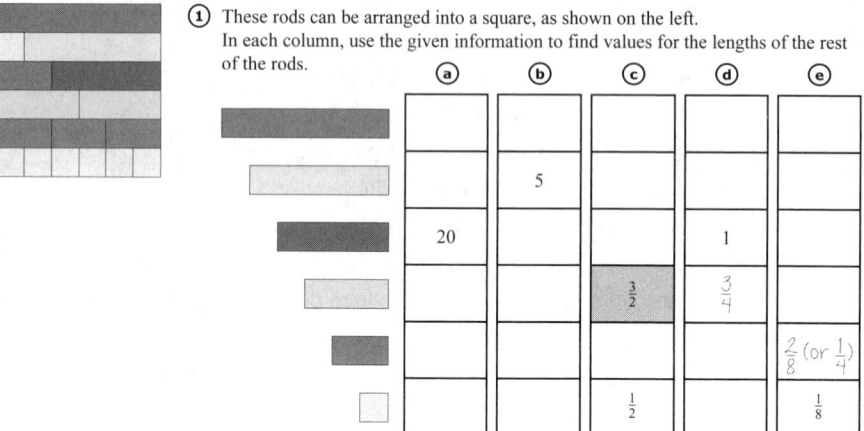

① These rods can be arranged into a square, as shown on the left.
In each column, use the given information to find values for the lengths of the rest of the rods.

	ⓐ	ⓑ	ⓒ	ⓓ	ⓔ
		5			
	20			1	
			$\frac{3}{2}$	$\frac{3}{4}$	
					$\frac{2}{8}$ (or $\frac{1}{4}$)
			$\frac{1}{2}$		$\frac{1}{8}$

After students have worked on most of the columns, have volunteers act out the *Thinking Out Loud* dialogue, using a copy of the table on page T39. Have students discuss the different ways of thinking about the number $\frac{3}{2}$. Use the *Pausing to Think* box to hear from students how they might make sense of the number $\frac{5}{2}$. The following question may also help guide the discussion about their observations:

» **What observations did you make about fractions as you filled in the table from problem 1?** Use this prompt to discuss some of the vertical or horizontal relationships in the table. For example:

• To fill out each column, it often helps to find the value of the white piece. Once you find the white piece, you can just "count" to fill in the rest of the column.

• The columns are related to each other. For example, every entry of column d is one-fourth of the corresponding entry in column b. Similarly, every entry of column c is one-tenth of the corresponding entry in column a. Because we're looking at relative lengths of rods, the lengths grow or shrink *proportionally*.

• $\frac{1}{8}$ is half of $\frac{1}{4}$, so every entry in column e is half of the corresponding entry in column d. From this, we can see that one way of finding half of a number is by multiplying the denominator by 2.

Student Problem Solving and Discussion

The problems in the Student Worktext ask students to use Cuisenaire rods and number lines to make sense of adding and subtracting fractions. Use the suggested discussion prompts to explore these ideas.

» **In problem 3, the two fractions you added had the same denominator. Use the number line and Cuisenaire rods to explain a general method for adding fractions that have the same denominator.** We can make sense of adding fractions using the number line by thinking of the denominator as defining the "units" of the problem. If every fraction is written in thirds, for example, then the sum of the fractions can also be located on a number line marked off in thirds. The same thinking applies to using Cuisenaire rods. If we know the value of the white piece (the "units"), adding blocks together becomes a matter of counting the number of white blocks it takes to make the total.

» **Problems 5–7 involve addition and subtraction of fractions with *unlike* denominators. You might have learned a rule for adding fractions that requires you to rewrite those fractions so that you are adding fractions that have the *same* denominators. How does your representation of each problem on the number line help you make sense of this rule?** Make sure that students understand that writing fractions in this way does not change the value of the quantities being added. Help them realize that when they visualize the problem $1 + \frac{2}{3}$ on a number line marked in thirds, the distance of 1 is simultaneously marked as "1" and "$\frac{3}{3}$," and in fact "$\frac{3}{3}$" is how a person might rewrite "1" in the problem to find a common denominator: $\frac{3}{3} + \frac{2}{3}$. That's what this "rule" does, and we can see visually why it is convenient to turn a problem like $1 + \frac{2}{3}$ into a problem where all the quantities are written in thirds.

» **How did you decide on the locations of the numbers in problem 11?** Listen for students who realized that $\frac{4}{3}$ and $\frac{8}{6}$ are equivalent, so they belong at the same location. Four is 3 times $\frac{4}{3}$, so 4 is 3 times farther from 0 than $\frac{4}{3}$. Just like $\frac{1}{6}$ is half of $\frac{1}{3}$, $\frac{8}{6}$ is half of $\frac{8}{3}$. And $\frac{3}{3}$ is equivalent to 1, so $\frac{4}{3}$ must be greater than 1.

» **All numbers have many names. What are some different names for the number 20? When might seeing the number 1 as $\frac{9}{9}$ help?** Some different names for 20 include $2 \cdot 10$, $5 \cdot 2 \cdot 2$, $16 + 4$, $\frac{40}{2}$, $100 - 80$, etc. Seeing the number 1 as $\frac{9}{9}$ might help when figuring out how large $\frac{10}{9}$ is, or finding $1 + \frac{7}{9}$, and so on.

Adding fractions with the same denominator is a lot like counting. 7 ninths plus 7 ninths will be 14 ninths, just as 7 apples plus 7 apples would be 14 apples. The denominator of the sum will be the same as the denominator of the addends, whether it's ninths or apples. The numerator of the sum will be the sum of the numerators of the addends. It's the distributive property again: 3 fives plus 4 fives is 7 fives: $3 \cdot 5 + 4 \cdot 5 = 7 \cdot 5$.

Although students won't be using calculators in this unit, and some modern calculators do allow fractions to be entered, it can be instructive to ask students how to enter a number like $4\frac{1}{2}$ into a calculator. Common incorrect responses include typing 4.2, $41 \div 2$, and $4 \times 1 \div 2$. Listen for students who understand that $4\frac{1}{2}$ is between 4 and 5, so the decimal representation for the number must be "a kind of 4." The target number is $\frac{1}{2}$ more than 4, so one correct input is $4 + (1 \div 2)$. Note that some calculators may be able to parse the expression $4 + 1 \div 2$ correctly as $4\frac{1}{2}$, but simpler models may treat it as $(4 + 1) \div 2$. Students may also suggest typing 4.5; it is great for students to see this connection. Ask students for a way to input $4\frac{1}{7}$ or $10\frac{5}{3}$ to make sure they understand that this notation implies addition.

Preparation: Prepare to display the "Cuisenaire Table Launch" page (available in the Resources PDF; a reference copy is provided on page T40).

Mental Mathematics (5 min)

Launch: Cuisenaire Table (15 min)

· Solve together and discuss students' reasoning about multiplication and division.

Student Problem Solving and Discussion (25 min)

· Allow students to work through the Important Stuff and explore additional problems.

· Discuss the strategies for multiplication with fractions addressed in the *Discuss & Write* box.

Unit 8 Related Activity: Making MysteryGrids (See page T37 and Student Worktext page 50.)

Lesson 2:
Multiplying with Fractions

PURPOSE

In this lesson, students deepen their understanding of the operations of multiplication and division to support sense-making in algebra. For example, in problem 11, students multiply numbers by 3. The numbers are written in ways that suggest using different strategies for multiplication. For example, in finding $\frac{8}{3} \cdot 3$, one might take advantage of the inverse relationship between dividing by 3 and multiplying by 3, whereas when multiplying $\frac{1}{17} \cdot 3$, one might think of "seventeenths" as a unit, and realize the product would be 3 seventeenths. By examining multiplication in these different ways, students get a better idea of what the expression $3x$ represents. By attending to these different approaches, students also learn to look for mathematically strategic shortcuts, which is an important part of seeking and using structure.

In a similar way, comparing the magnitudes of quantities such as $\frac{1}{3} \cdot 3$ and $\frac{1}{6} \cdot 3$ helps students understand the relationship between $\frac{1}{3}x$ and $\frac{1}{6}x$. Understanding how these magnitudes compare for positive and negative values of x will support what students will see when they compare the graphs of these expressions in Unit 9: *Points, Slopes, and Lines*.

The goal is for students to understand multiplication of fractions in a logical and flexible way. In Lesson 2, sets of problems highlight ways of looking at multiplication of fractions by integers. Lesson 3 will address multiplying fractions by fractions. Why this progression? The focus on multiplying integers develops ideas from Lesson 1, using the logic of adding fractions with like denominators. Also, multiplication by integers gets students very familiar with the idea that $a \cdot \frac{b}{a} = b$. Students will extend this idea in several ways in Unit 8. For example, students who are comfortable with the calculation $\frac{3}{5} \cdot 5 = 3$ can use that to make sense of the calculation $\frac{3}{5} \cdot \frac{5}{4} = \frac{3}{4}$. Students will also use this idea when they solve equations such as $\frac{7}{2}x = 6$ by doubling the relationship to get $7x = 12$ as part of their solving process.

Mental Mathematics Begin each day with five minutes of Mental Mathematics (pages T51–T64). These activities help students articulate and formalize important ideas about fractions.

Launch: Cuisenaire Table

Display "Cuisenaire Table Launch" (page T40). Give students time to look over the entire table in silence. Then ask students to come forward to fill in just one box at a time and explain their reasoning. The solution is shown here (though any box can be filled in sensibly with an equivalent fraction, depending on the reasoning students use).

	ⓐ	ⓑ	ⓒ	ⓓ	ⓔ
	6	18	9	$\frac{18}{5}$	$\frac{9}{5}$
	5	15	$\frac{15}{2}$	3	$\frac{3}{2}$
	4	12	6	$\frac{12}{5}$	$\frac{6}{5}$
	3	9	$\frac{9}{2}$	$\frac{9}{5}$	$\frac{9}{10}$
	2	6	3	$\frac{6}{5}$	$\frac{3}{5}$
	1	3	$\frac{3}{2}$	$\frac{3}{5}$	$\frac{3}{10}$

The table is designed so that students can compare the columns. For example, column b is 3 times a, column c is half of b, column d is a fifth of b, and column e is half of d. These patterns provide opportunities for discussion about multiplication and division. Below are some possible discussion questions. Responses will vary depending on how students have filled in the boxes.

» **We can compare rows in this table. Compare the numbers in the two different green rows. How are these numbers related?** The light green rod is half the length of the dark green rod. So the numbers in the light green row are half of the numbers in the dark green row. Students can see different ways of writing half of a number: half of 9 is $\frac{9}{2}$, half of $\frac{18}{5}$ is $\frac{9}{5}$, and half of $\frac{9}{5}$ is $\frac{9}{10}$.

» **The numbers in column e are half of the numbers in column d. How can we see this relationship?** With the fractions written as above, there are three ways that we see division by 2 written. Half of 3 is $\frac{3}{2}$; in this case, we show division by writing a fraction. Half of $\frac{6}{5}$ is $\frac{3}{5}$; in this case (and in the case of halving $\frac{12}{5}$ and $\frac{18}{5}$) *the numerator has been halved*. Half of six-fifths is three-fifths. Finally, we see half of $\frac{3}{5}$ is $\frac{3}{10}$; in this case (and in the case of halving $\frac{9}{5}$) *the denominator has been doubled*. A tenth is half of a fifth (we can picture this on a number line from 0 to 1, since marking tenths requires twice as many subdivisions), so half of 3 fifths must be 3 tenths.

» **The numbers in column c are larger than the numbers in column a. How many times larger?** By looking at the relative measurements of the white rod, we see that the numbers in column c are $\frac{3}{2}$ times larger than the numbers in column a. To make sense of this, consider taking each number in column a and multiplying by 3 and dividing by 2. (The intermediate calculation is actually shown in column b.) So column c is the result of performing two operations on column a: multiplying by 3 and dividing by 2. These can be written as one operation: multiplying by $\frac{3}{2}$.

Many other relationships can be discussed, but this brief start is meant to address a few ideas about multiplication and division and demonstrate ways of identifying patterns that help make sense of these operations and will support later work with proportional relationships. Students will examine these operations in more detail throughout this lesson and the next few lessons.

Student Problem Solving and Discussion

Give students time to work on the Important Stuff and additional problems on their own or in groups.

Students work through several sequences of problems intended to highlight patterns that they can use to figure out what happens when a fraction is multiplied by an integer. In several cases, students also express these patterns using algebra.

Ask students to share their responses for **PROBLEM 12** from the *Discuss & Write* box.

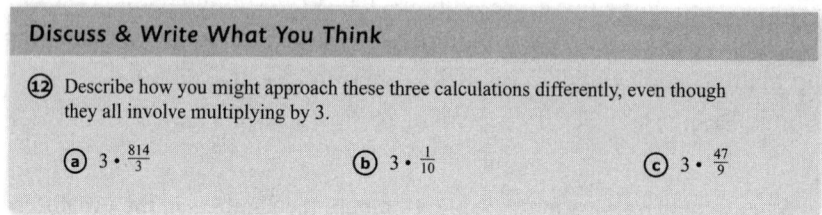

> **Discuss & Write What You Think**
>
> ⑫ Describe how you might approach these three calculations differently, even though they all involve multiplying by 3.
>
> ⓐ $3 \cdot \frac{814}{3}$　　　　ⓑ $3 \cdot \frac{1}{10}$　　　　ⓒ $3 \cdot \frac{47}{9}$

Ask students to verbalize *why* these methods work. For example, to describe why $3 \cdot \frac{814}{3} = 814$, a student might observe that 814 is both being divided by 3 and multiplied by 3 in the problem, and those operations undo each other. To describe why $3 \cdot \frac{1}{10} = \frac{3}{10}$, a student might refer to a number line image to explain that tripling a length of *one* tenth produces a length of *three* tenths. To describe why $3 \cdot \frac{47}{9} = \frac{47}{3}$, students might describe how thirds are three times as long as ninths. They can use a number line from 0 to 1 to show this. It follows that 47 thirds will be three times as long as 47 ninths. Students might use other lines of reasoning to describe their strategies. Students may also express their products differently. For example, a student who says $3 \cdot \frac{47}{9} = \frac{141}{9}$ is correct and likely has a valid strategy for multiplying fractions by 3.

Use these prompts to elicit student thinking about multiplication and division:

> » **In problem 2, how did you make sense of the fact that $\frac{3}{4} \cdot 4$ is equivalent to $4 \div 4 \cdot 3$?** This equivalence might be especially challenging for students confused about the order of operations for multiplication and division. Students may find it helps to realize that $\frac{3}{4} \cdot 4$ is equivalent to $4 \cdot \frac{3}{4}$. Multiplying any quantity by $\frac{3}{4}$ involves both multiplying by 3 and dividing by 4 (in either order).

> » **In problem 16, only one of the responses was false. Which one? Explain.** Use this prompt to highlight the idea that multiplying by 5 is different from

---

Students may have trouble with order of operations in calculations like 4 ÷ 4 • 3 or 2 – 5 + 3, concluding, incorrectly, that these are equivalent to 4 ÷ 12 and 2 – 8. This may stem from applying a mnemonic, such as PEMDAS (parentheses, exponents, multiplication, division, addition, subtraction), without understanding the underlying meaning. Because inverse operations (addition/subtraction and multiplication/division) do not have a special order and should instead be considered in the order written on the page, some teachers use the following representation:

P

E

M/D

A/S

Students do not need to learn a new mnemonic, but the underlying meaning, of course, is essential.

multiplying by $\frac{5}{5}$. Multiplying by $\frac{5}{5}$ is the same as multiplying by 1 and results in a number equivalent to the original value.

» Write the mobile in **PROBLEMS 17–19** as an equation: $4s = 3t$ (where s = the weight of the square and t = the weight of the triangle). Plug in the values for the different problems. For example, plugging in values in problem 19 will yield the equality $4 \cdot \frac{1}{4} = 3 \cdot \frac{1}{3}$, which students can discuss. **In each case, which value is larger, the weight of the square or the weight of the triangle?** Students can make sense of the fact that the weight of the triangle is always larger than that of the square because a larger number of squares balances a smaller number of triangles. Students will examine this idea in further detail in Lesson 7.

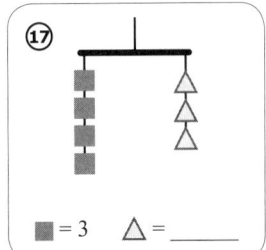

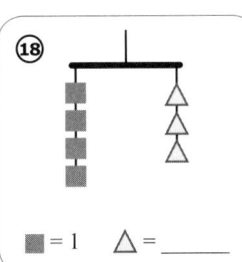

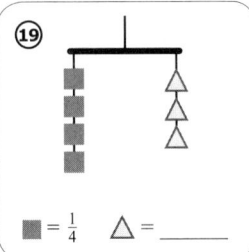

What if . . .

What if students don't simplify their fractions?

Simply, that's fine. Many of the solutions provided in the Answer Key are written in reduced form (i.e., the greatest common factor of the numerator and denominator is 1). This is mostly a practical consideration, because it is impossible to include the infinitely many forms of the solution to any problem. Occasionally, non-reduced fractions are also included in the Answer Key as a reminder that other responses are acceptable and correct. Use the fact that students are generating different forms of the same solution as an opportunity to help students *build reasoning*. Students who write $\frac{1}{4} \cdot 4 = \frac{4}{4}$ are completely correct, and their understanding of multiplication might provide a flash of insight to other students who now see an important connection between $\frac{4}{4}$ and 1. *The most useful form of a fraction depends on the problem situation and the chosen solution strategy.* That means that when demonstrating a particular strategy for students, it might be sensible to give preference to writing fractions in the form that is most helpful to the specific context (often *unreduced*). But in general, with student-generated solutions, affirm the reasoning students use to arrive at their form of a solution.

Mental Mathematics (5 min)

Launch: *Thinking Out Loud* **Dialogue**
(20 min)

· Give time for students to do problems 1–3.

· Have students act out the dialogue.

Student Problem Solving and Discussion
(20 min)

· Give students time to work through the Important Stuff and explore additional problems.

· Share responses from the *Discuss & Write* box about how the area model shows the process of multiplying fractions.

Unit 8 Related Activity: Making MysteryGrids (See page T37 and Student Worktext page 50.)

Lesson 3:
Fractions and Area Models

PURPOSE

The purpose of Lesson 3 is for students to make sense of the multiplication of fractions and show that it behaves in a way that is consistent with the multiplication of integers. Students use area models for fractions as they have done before with integers and algebraic expressions. The context of area helps students visualize multiplication by a non-integer quantity. Students also use the area model representation to make clear why one can find the product of two fractions by multiplying their numerators and denominators $\frac{a}{b} \cdot \frac{c}{d} = \frac{ac}{bd}$.

Mental Mathematics Begin each day with five minutes of Mental Mathematics (pages T51–T64). As students continue to gain experience with fractions, encourage them to imagine models, such as the number line or area models, to help them visualize their calculations.

Launch: *Thinking Out Loud* Dialogue

Provide time for students to work through **PROBLEMS 1–3** on their own, individually or in small groups. After they are done, students should read through the dialogue silently and consider the *Pausing to Think* prompts.

Have three volunteers play the characters in front of the class. Where the dialogue calls for drawing models and making markings, your student actors should draw the models and make all the indicated markings on the board for all to see. Ask students where the denominator in the product, 21, appears in the diagram (this is the number of identical pieces that are cut in the unit square) and how that number arises (from splitting up the sides into 3 and 7 partitions). Then ask where the numerator of the product, 6, appears in the diagram (it's the number of twenty-firsts that are being counted by the calculation) and how that number arises (from counting the pieces in a rectangular region that has 1 of those pieces along one side and 6 along the other). Then, discuss the *Pausing to Think* prompt. Students should recognize that as long as the 1×1 square has been partitioned into 21 equal spaces, shading *any* 6 of those spaces represents the fraction $\frac{6}{21}$. Students may also recognize that $\frac{6}{7}$ of the 1×1 square can be represented as the shaded area shown to the right, so $\frac{1}{3} \cdot \frac{6}{7}$ is a third of the shaded area, which is $\frac{2}{7}$ (two of the rows that each represent one seventh). This is one way of looking at the equivalence of $\frac{6}{21}$ and $\frac{2}{7}$.

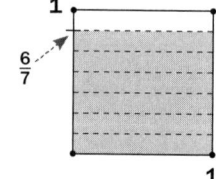

You may wish to demonstrate the multiplication process with another example to emphasize why multiplying fractions across works. Consider using $\frac{2}{3} \cdot \frac{7}{10}$ and again asking students to demonstrate where the 30 in the denominator of the product and the 14 in the numerator of the product appear in the diagram and what aspects of the diagram generate these numbers. With this example, the equivalence to $\frac{7}{15}$ is not as straightforward to show. However, once students agree that $\frac{2}{30} = \frac{1}{15}$, they can outline adjoining pairs of thirtieths and count that there are, indeed, 15 of these regions with equal area (though different shapes) and that exactly 7 of them are counted by the calculation.

Student Problem Solving and Discussion

Students should work on the rest of the Important Stuff and explore additional problems. **PROBLEMS 4–6** help students consider the many ways that fractions might be represented on an area model. **PROBLEMS 7–13** provide experience with multiplying fractions using area models.

Use **PROBLEM 16** in the *Discuss & Write* box to encourage students to review how using area models justifies the process of multiplying fractions by multiplying. Listen for students who appreciate having a visual image that confirms and validates this method. **PROBLEMS 17 & 18** examine patterns in multiplication and include multiplying fractions with variables. Even though students have a method for multiplying fractions that always works, looking for patterns and considering related problems can often make calculations simpler.

Use these prompts to further discuss multiplication of fractions:

» **Show how you might use a number line to model the problem $\frac{1}{2} \cdot 6$. How could you model the same calculation with an area model?** The problem can be thought of as finding "half of 6" or "6 pieces of length (or area) $\frac{1}{2}$." Invite students to draw number lines and area models. Some possible drawings are shown here.

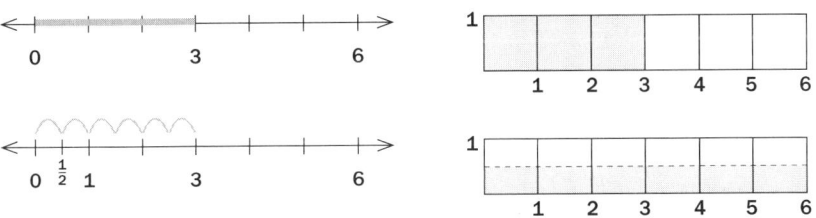

» **What was the pattern in each of problems 17 and 18 that helped make the calculations simpler?** In problem 17, each calculation used the fact that $8 \cdot \frac{1}{4}$ is 2. This leads to at least two ways to think of $\frac{8}{3} \cdot \frac{1}{4}$. One way is to consider that it involves an extra division by 3, so the product is $\frac{2}{3}$. Another way is to "hear" the fraction $\frac{8}{3}$ as "8 thirds." A quarter of 8 thirds is 2 thirds. In problem 18, each calculation allows us to divide a multiple of 3 by 3 to simplify the calculation.

» **Is the product of two positive fractions always smaller than either factor? Explain.** The size of the product depends on the size of the factors, not on whether the factors are represented as fractions or in some other way. To reinforce the idea that the product is less than either factor only when the factors are between 0 and 1, emphasize that the first factor is already less than 1, which is represented by the unit square, and the second factor reduces this area. You can also show an extra intermediate step when drawing area models. For example, demonstrating the multiplication $\frac{2}{5} \cdot \frac{1}{3}$ (problem 8) in the following two steps would help students see how the area is getting smaller.

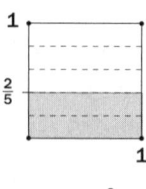
Area: $\frac{2}{5}$

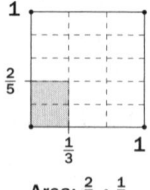
Area: $\frac{2}{5} \cdot \frac{1}{3}$

Students may also consider what happens when the factors are both greater than 1 (the product is always greater than either factor), one factor is greater than 1 and the other is between 0 and 1 (the product can be either greater than or less than 1), and either or both of the factors are negative.

Lesson 4:
Rewriting Rational Expressions

PURPOSE

The expressions $\frac{1}{5}(10x + 15)$ and $\frac{10x + 15}{5}$ represent the same underlying computation; given a value for x, both produce the same result. But they look very different. One looks like multiplication, and the other looks like division; one contains a fraction, and the other *is* a fraction; and yet they are equivalent. An overall goal of Unit 8, and the special purpose of Lesson 4, is to help students use this interplay of ideas to build one coherent image relating multiplication, division, and rational expressions.

In Lesson 4, students use area models to rewrite algebraic expressions that involve fractions. Students also encounter the idea that area models can be used to represent division when used "inside out." This important idea will return in Unit 10: *Area Model Factoring*.

Mental Mathematics Begin each day with five minutes of Mental Mathematics (pages T51–T64). These activities support the idea that different-looking calculations (e.g. tripling and finding the number of thirds in something) can amount to the same thing.

Launch: Comparing Models

Display the top half of "Comparing Models" (page T41). Ask students to explain how each of the models shows $\frac{4}{3}$.

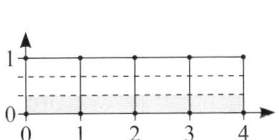 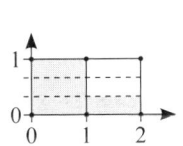 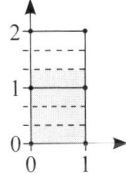

Students may notice that all of the smaller rectangles have an area of $\frac{1}{3}$ and that four of these rectangles are shaded in each figure. Listen for students who also notice differences in how the models show the quantity. For example, the first model supports the idea that $\frac{4}{3}$ is a third of 4 (one-third of the entire 1×4 area) and that it is 4 times $\frac{1}{3}$ (a rectangle with area $4 \cdot \frac{1}{3}$; or, alternatively, 4 times the $\frac{1}{3}$ rectangle). The second and third models are two ways of showing that $\frac{4}{3}$ is equivalent to $1\frac{1}{3}$. The third model shows a green rectangle with area $1 \cdot \frac{4}{3}$.

Then ask students to use the models to show the quantity $\frac{1}{2} \cdot \frac{4}{3}$. Students may represent the quantity by indicating two of the four sections to represent $\frac{2}{3}$. Students may also represent this operation by splitting each of the smaller rectangles into two equal parts to make sixths. Some of the possible

Lesson at a Glance

Preparation
- Photocopy the Snapshot Check-in on page T45.
- Prepare to display "Comparing Models" (available at in the Resources PDF; a reference copy is provided on page T41) for the Launch.

Mental Mathematics (5 min)

Launch: Comparing Models (10 min)
- Students compare models that show $\frac{4}{3}$ and discuss different ways of showing half of $\frac{4}{3}$.

Student Problem Solving and Discussion (20 min)
- Have students work through the rest of the Important Stuff and explore additional problems.
- Discuss equivalent ways of writing rational expressions.

Reflection and Assessment: (10 min)
Snapshot Check-in

Unit 8 Related Activity: Making MysteryGrids (See page T37 and Student Worktext page 50.)

representations are shown at the bottom of "Comparing Models" (others are possible). Students should discuss how all these ways of showing $\frac{1}{2} \cdot \frac{4}{3}$ are equivalent and show the quantity $\frac{2}{3}$. For each area model that students discuss, create a corresponding expression and discuss its equivalence to $\frac{1}{2} \cdot \frac{4}{3}$.

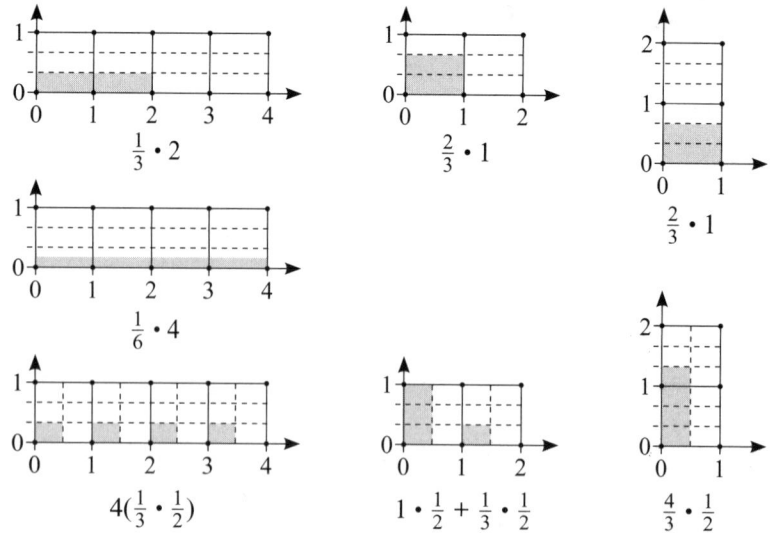

$\frac{1}{3} \cdot 2$

$\frac{2}{3} \cdot 1$

$\frac{2}{3} \cdot 1$

$\frac{1}{6} \cdot 4$

$4\left(\frac{1}{3} \cdot \frac{1}{2}\right)$

$1 \cdot \frac{1}{2} + \frac{1}{3} \cdot \frac{1}{2}$

$\frac{4}{3} \cdot \frac{1}{2}$

❓ What if...

What if students are confused by the use of division with the area models?

Because expressions that involve division can be written as multiplication (and vice versa), e.g. $\frac{x+3}{2} = \frac{1}{2}(x+3)$, a single rational expression can be represented by different area models. For example, both of the area models below may be used for either of the expressions in the equation given above. The first shows multiplication, and the second, division:

	$\frac{x}{2}$	$\frac{3}{2}$
2	x	3

$$\frac{x+3}{2} = \frac{x}{2} + \frac{3}{2}$$

	x	3
$\frac{1}{2}$	$\frac{x}{2}$	$\frac{3}{2}$

$$\frac{1}{2}(x+3) = \frac{x}{2} + \frac{3}{2}$$

It is important that students recognize what is consistent in all area models: that the two side lengths are always multiplied to give the product (the area) within and, correspondingly, the area inside can be divided by an expression written along one of its sides and this will result in the other side length.

Note that while area models are an excellent tool for organizing multiplication, they are not always convenient for organizing arbitrary division problems, e.g. $\frac{1}{x-1}$. But whenever an equation can be expressed as two finite polynomial factors and their product, the area model can represent it as multiplication, division, or factoring, greatly helping to explain the relationships among those operations. Students will address these ideas in greater depth in Unit 10: *Area Model Factoring*.

Student Problem Solving and Discussion

PROBLEM 1 asks students to identify expressions equivalent to half of 28. **PROBLEMS 2–5** and **6–9** are two similar sequences of area model problems that start with integers and lead to fractions. **PROBLEMS 12–17** require students to examine the structure of the expression and represent that structure by filling in the area model to show multiplication or division. Note that some of the blank spaces on the area models in problems 12–14 are filled in to draw attention to how the area model may be set up to show either multiplication or division. In later problems, however, students fill in the area model in the way that makes most sense to them. For example, though the expression in problem 15 shows division, $\frac{35c - 2d + 50}{5}$, a student may choose to fill in the area model to find the product of $\frac{1}{5}$ and $35c - 2d + 50$. Both are correct.

The area models in **PROBLEMS 21–24** require students to perform multiplication and division tasks that require more thought. For example, in problem 23, the first blank space in the area model asks students to find $4 \cdot \underline{\hspace{1em}} = 7h$. Remind students of patterns they noticed in Lesson 2 when they solved similar problems.

Use these questions to encourage students to consider what is important when writing equivalent expressions:

» **Problem 1 asks you to select the expressions that are equivalent to half of 28. Explain what is wrong with the expressions you didn't circle.** Choices C and J show finding half of only 20, then adding 8. Choice F, $28\frac{1}{2}$, is the number halfway between 28 and 29; it is a way of writing $28 + \frac{1}{2}$, not $28 \cdot \frac{1}{2}$. Choice E, $\frac{2}{2} \cdot 28$, is *two* halves of 28; it is equivalent to 28. And choice G, $\frac{1}{2 \cdot (20+8)}$, puts the 28 in the denominator; evaluating the $2 \cdot (20 + 8)$, we see that this fraction is $\frac{1}{56}$, *much* smaller than half of 28.

» **Problem 20 also asks you to identify equivalent expressions. This time, it was for expressions equivalent to $\frac{12x-6}{3}$. Explain what is wrong with the expressions you didn't circle.** Choice C, $\frac{6-12x}{3}$, is the negative; the signs in the numerator are opposite from the original expression. Choice D, $4x-6$, shows a third of $12x$, but not of 6. Students distracted by Choice G, $\frac{6x}{3}$, might incorrectly think that $12x-6=6x$.

» **In problems 10 and 11 we see that when an expression is written as a fraction, one way of representing it in an area model is to put the numerator on the *inside* of the area model. Why does it make sense to write this expression on the inside of the model rather than the outside?** Multiplication and division are inverse operations. They undo each other. If multiplying the length and width of a rectangle gives its area, then to undo this process, one can start with the area of the rectangle and divide by the length to get the width (or divide by the width to get the length). Because we think about the area of a rectangle as a product of its dimensions, it makes sense that a division problem represents the quantity being divided (the numerator, if the division is written as a fraction) as the area, the space inside of the rectangle.

» **There are many ways to write equivalent expressions for problems 12–17. What are some of these expressions?** For example, in problem 12, listen for students to express the idea that multiplying by $\frac{1}{8}$ is the same as dividing by 8. Students should see how this thinking translates to the area model: *multiplying* (in this case, by $\frac{1}{8}$) is represented by writing both factors along the two outside side lengths of the two-dimensional model with the goal of finding the area inside (the product); *dividing*, on the other hand, starts with the area inside (the number being divided) and one of the side lengths (the divisor, in this case, 8), with the goal finding the other side length of the rectangle.

» **Express $\frac{x+3}{2}$ with an area model. Then express $\frac{3}{x+2}$ with an area model. Discuss the difference between what these two expressions mean. Use area models to show that $\frac{x+3}{2}=\frac{x}{2}+\frac{3}{2}$, but $\frac{3}{x+2}$ does not equal $\frac{3}{x}+\frac{3}{2}$.** With the expression $\frac{x+3}{2}$, the area model shows how it is possible to think of the expression as half of x plus half of 3—especially if you sketch a horizontal line through the middle of the model. The second expression, $\frac{3}{x+2}$, cannot be thought of in the same way; we cannot break up 3 into pieces proportionate to the sum of x and 2 because we don't know the value of x. To demonstrate that $\frac{3}{x+2}$ does *not* equal $\frac{3}{x}+\frac{3}{2}$, students can use an area model to show that $(x+2)(\frac{3}{x}+\frac{3}{2})$ does not equal 3. (Make sure that students understand why this approach is a valid way to show that $\frac{3}{x+2}$ and $\frac{3}{x}+\frac{3}{2}$ are not equivalent.)

Student Reflections &
Snapshot Check-in

Ask students to reflect on their learning:

What are some things you've learned so far in this unit?

What questions do you still have?

Assess student understanding of the ideas presented so far in the unit with the Snapshot Check-in on page T45. Use student performance on this assessment to guide students to select targeted Additional Practice problems from this or prior lessons as necessary.

So far in Unit 8, students have:

- Explored fractions using Cuisenaire rods of various lengths, mobiles, the number line, and area models.
- Found different approaches for multiplying fractions by integers.
- Represented multiplication of fractions using area models and connected the image to a general method for multiplying fractions.
- Written and identified equivalent rational expressions for both numerical and algebraic rational expressions.

Students have also focused on the following Algebraic Habits of Mind:

- **Using Tools Strategically**—In applying the number line and area models to understanding fractions, students have practiced using tools to organize and explain their thinking and consider the reasonableness of an answer. By adapting area models and extending their use to fractions, students have gained new insight about fraction multiplication.
- **Seeking and Using Structure**—Students have seen that working with fractions often requires thinking about them in more than one way. Students seek and use structure when they consider equivalent forms of rational expressions and use a form they find convenient to approach a problem.

Lesson 5:
Scaling to Solve

PURPOSE

Understanding equality is essential to understanding algebra. Many students are already comfortable with the idea that dividing both sides of an equation by the same (non-zero) number maintains equality. In equations with integer coefficients (as in mobiles), it is less obvious why anyone would ever want to *multiply* both sides. But it does become useful sometimes when an equation contains fractions. Students have already seen that multiplying a fraction by its denominator results in an integer. This method is more than just about "getting rid of a fraction," however. It is about understanding how multiplication affects a balanced equation and taking advantage of this understanding to make sense of equations with fractions. Using this idea, students can approach equations with fractions using the logic and general methods that they would apply in solving any other equations. For students to understand this method, they must have a mental image of how quantities are affected by scaling. Students use number lines to compare quantities like $\frac{2}{5}x$ and x in order to make sense of these relative quantities.

 ***Mental Mathematics*** *Begin each day with five minutes of Mental Mathematics (pages T51–T64). Activities like "Approximation" and "Closer approximation" give students a sense of what constitutes a reasonable answer when dividing. These activities encourage students to estimate instead of immediately applying an algorithm.*

Lesson at a Glance

Mental Mathematics (5 min)

Launch: *Thinking Out Loud* Dialogue (15 min)

· Give time for students to do problems 1–4, then have students prepare and act out the dialogue.

· Students consider how multiplying both sides of an equation can help in finding a solution.

Student Problem Solving and Discussion (25 min)

· Provide time for students to work through the Important Stuff and additional problems.

· Ask students to share their responses from the *Discuss & Write* box, which connects the scaling method of solving equations to the number line.

· Discuss how forming a mental image of a relationship can help in solving equations.

Unit 8 Related Activity: Making MysteryGrids (See page T37 and Student Worktext page 50.)

Launch: *Thinking Out Loud* Dialogue

Give students time to work through **PROBLEMS 1–4** on their own, individually or in small groups. Have them read the *Thinking Out Loud* dialogue silently, and ask three volunteers to act out the dialogue.

As a class, address the *Pausing to Think* boxes. The first box asks students to find the solution, $x = 12$. After finding this solution, ask students to consider its reasonableness: **Is it reasonable that 12 is the answer to the question "Five-thirds of what number is 20?" Did you expect the answer to be bigger or smaller than 20?** The second box asks students to consider Michael's last line in the dialogue. Ask students to reason through solving the original equation, $\frac{5}{3}x = 20$, using Michael's proposed first step of dividing both sides by 5. All of these methods lead to the same solution, even though the calculations are quite different. Students should see that both multiplying both sides by 3 and dividing both sides by 5 change the numbers in the equation while still keeping the relationship balanced.

Use ideas of *scaling* (such as "We multiply by 4 to make both sides of the equation 4 times larger") and *relative position* on the number line (such as "Do we expect $\frac{1}{3}x$ to be larger or smaller than x?") rather than cross multiplication. Cross multiplication is a common source of errors, as it is often misapplied as a technique for multiplication. If students mention cross multiplication, acknowledge the relevance of their idea, and say that the lesson's focus is on understanding moves that keep an equation balanced, and this will give the same result, but in a clearer way than using cross multiplication. Some students may also be ready to discuss the connection to cross multiplication: "We know we have to do the same thing to both sides of an equation to keep it balanced. How is cross multiplication related to what we have been doing today?"

Student Problem Solving and Discussion

PROBLEM 5 leads students through various ways of scaling an equation.

⑤ Here are three different ways to scale the equation $\frac{2}{3}x = 7$. Finish solving for x in each case.

ⓐ $3 \cdot \frac{2}{3}x = 7 \cdot 3$ ⓑ $\frac{1}{2} \cdot \frac{2}{3}x = 7 \cdot \frac{1}{2}$ ⓒ $\frac{3}{2} \cdot \frac{2}{3}x = 7 \cdot \frac{3}{2}$

$\frac{1}{3}x = \frac{7}{2}$

PROBLEM 6 asks students to solve equations by finding their own sensible way of scaling the equation.

⑥ Solve for x by finding a sensible way to scale the equation.

ⓐ $\frac{5}{2}x = 25$ ⓑ $2 = \frac{3}{8}x$ ⓒ $\frac{1}{3}x = \frac{1}{2}$

For many students, it will make sense to talk about "getting rid of the fraction" as a way of entering into the problem. In your conversations with students, however, use language that emphasizes how multiplication and division are being used to scale the equations. For example, in problem 6a, talk about how the first step in solving $\frac{5}{2}x = 25$ might be to double both sides of the equation. This kind of language will help students keep thinking of how equality is maintained as they solve these equations. "Getting rid of the fraction" is also not necessarily the goal of a problem; for example, in problem 6c, students solve $\frac{1}{3}x = \frac{1}{2}$. Students who are only focused on "getting rid" of fractions will find that they are unable to get rid of them all, as the solution itself is a fraction.

PROBLEMS 7–14 ask students to solve equations by using a number line to model each problem. Being able to represent the problem visually in this way is not meant to be a solution technique, but rather a tool students can use to judge the reasonableness of their solutions. In particular, **PROBLEMS 14 & 15** ask students to analyze a problem without going through the calculations to solve the problem.

⑭ If $\frac{5}{2}x = 15$, is x greater or less than 15?   ⑮ If $\frac{2}{3}x = \frac{5}{8}$, is x greater or less than $\frac{5}{8}$?

As a class, have students share their responses for problem 7 in the *Discuss & Write* box, which asks them to use the number line to solve the equation $\frac{5}{3}x = 20$. In particular, listen for students who understand that locating $\frac{5}{3}x$ and 20 at the same position on the number line shows their equality. Likewise, finding the number located at the same position as x is equivalent to solving for x.

You might use this prompt to start a discussion about forming a mental picture to make sense of equations:

> » **Share your thinking for problems 14 and 15. How would you represent each of these equations using a number line?** Encourage students to compare these problems to more familiar equations, such as $3x = 15$. Students will likely be able to identify that x must be smaller than 15 in this case, because the value of x must be tripled to get 15. Apply this reasoning to problem 14: if $\frac{5}{2}x = 15$, x must be less than 15 because it must be more than doubled to equal 15. A number line for each problem is shown in the Answer Key.

Lesson 6:
Scaling to Add

PURPOSE

Students continue to use reasoning they have developed about fractions to make sense of fraction addition. Again, the purpose is not to prescribe a *method* for adding fractions, but instead to lead students to a sensible approach to such problems. They use what they know about the size of fractions to predict where on a number line a sum will fall. They use what they have learned about equivalence to rewrite expressions and manipulate equations. Ultimately, this lesson is about using logic to approach unfamiliar problems.

 **Mental Mathematics** Begin each day with five minutes of Mental Mathematics (pages T51–T64). In the last few activities of the unit, students identify and work with factors of numbers.

Launch: *Thinking Out Loud* Dialogue

Give time for students to read through the dialogue silently. Ask for three volunteers to act out the dialogue in front of the class, writing the equations on the board where indicated. Address the three *Pausing to Think* prompts as a class.

Student Problem Solving and Discussion

Students who know the traditional algorithm for adding fractions by rewriting them with common denominators should be encouraged to use the familiar method. To make sure they are making sense of why this algorithm works, ask questions about the process that get them to describe the logic behind the steps they take.

> » **Why do you need a common denominator to add two fractions?** This question helps students reflect on the purpose and function of a common denominator. Listen for students who identify that a common denominator is a way of counting the same thing; $\frac{3}{8}$ and $\frac{1}{3}$ are three of one thing (eighths) and one of another (thirds)—they cannot be combined by adding 3 and 1. Using a common denominator of 24 allows us to find fractions that are equivalent to the original two ($\frac{9}{24}$ and $\frac{8}{24}$). Because these fractions count the same thing (twenty-fourths), we can add them together.

Lesson at a Glance

Mental Mathematics (5 min)

Launch: *Thinking Out Loud* Dialogue (15 min)

· Have students read and act out this dialogue, in which the characters make sense of a sum of fractions.

Student Problem Solving and Discussion (25 min)

· Give students time to work through the Important Stuff and additional problems.

· Discuss the logic behind different ways of manipulating an equation.

Unit 8 Related Activity: Making MysteryGrids (See page T37 and Student Worktext page 50.)

» **Where is $\frac{3}{8} + \frac{1}{3}$ on the number line? Where is $\frac{9}{24} + \frac{8}{24}$ on the number line?** Make sure that students understand that the method of finding equivalent fractions *doesn't* change the quantities. This is different from what the characters in the dialogue were doing. Using the method in the dialogue, the characters would start with $\frac{3}{8} + \frac{1}{3} = x$ and multiply the equation by 8 and 3 to make both sides of the equation 24 times larger. The value of the original sum, x, stays the same, but the values in the sum on the left-hand side of the equation change. In contrast, rewriting both fractions in twenty-fourths keeps the value of the sum shown exactly the same.

Discuss **PROBLEMS 3 & 9** from the *Discuss & Write* boxes as a class. Problem 3 asks students to express precisely what it means to "change" an equation. In discussing why $\frac{1}{3} + \frac{5}{6} = w$ can be rewritten as $\frac{2}{6} + \frac{5}{6} = w$, remind students of work they have done with mobiles, particularly in Unit 5: *Logic of Algebra,* by talking about what moves keep an equation balanced. In this case a "change" made to one side of the equation is *not* made on the other side. Discuss the idea that despite appearances, the equation has not fundamentally changed; all that changed was how part of one side was written. Students have encountered the same logic in solving mobile puzzles: they know that moves such as rearranging and substituting may affect only one side of a mobile or equation but can still be helpful as long as they don't change the balance. Listen for students who recognize that subtle difference: "When you change one side of the equation, you have to change the other" is not quite right; the more precise statement uses "change the *value* of" rather than just "change." Replacing one quantity with an equivalent quantity does not affect the balance of the equation or mobile; the equality still holds.

Use these prompts to discuss the logic of manipulating equations:

» **Why doesn't it make sense to add fractions by adding the numerators and adding the denominators straight across?** Students can talk about why this method doesn't make sense by talking about how numerators and denominators play different roles. Denominators can be thought of as the "units" of the addition problem; for example, $\frac{5}{3} + \frac{2}{3}$ can be thought of as the sum of 5 thirds and 2 thirds, the sum of which is 7 thirds.

» **Is the equation $x + y = 4$ really equivalent to $2x + 2y = 8$ even though one equation, if thought of as a mobile, is "heavier" than the other?** These are not the same equation; it would not make sense to use the same mobile to represent them. But they are equivalent; they show the same relationship, so *every* value for x and y that work in one equation will work in the other equation. The same is true of the equivalent equations $x + y + 2 = 6$ and $x = 4 - y$ as well. Since these equations all represent the same relationship, it is possible to choose the equation that is most convenient for your purposes.

Lesson 7:
Proportional Reasoning

PURPOSE

This lesson builds understanding of proportional relationships. An equation for a proportional relationship is structurally the same as an equation relating two equivalent fractions. Students take advantage of this similarity as they apply the strategy of repeating and generalizing that they used in Unit 7: *Thinking Things Through Thoroughly*. Using this general problem-solving strategy helps students build confidence about the equations they write because going through the process of writing an equation requires that they first make sense of the quantities they are relating by using numerical examples.

 Mental Mathematics *Begin each day with five minutes of Mental Mathematics (pages T51–T64). These activities continue to familiarize students with factors. This helps students develop the habit of looking at numbers as products of other numbers.*

Launch: Mobiles and *Thinking Out Loud* Dialogue

Have students work on **PROBLEM 1**, in which they create the "simplest" mobile to show a proportional relationship between the two shapes.

After students have built their mobiles, ask them to share their observations about building these mobiles. Listen for students who noticed that, for example, if ■ = 2 and ▲ = 3, then three squares will balance two triangles in the mobile. The number of shapes needed seems to come from the weight of the *other* shape. Interesting observation! Is that always the case? This is a place for students to perform other numerical experiments and see if they can build a real conjecture. What if we try ■ = 4 and ▲ = 6? What if we try ■ = 6 and ▲ = 9? What if we try ■ = 6 and ▲ = 12? As it happens, in each case, we've made the square the lighter of the two. Students might observe that it makes sense that there would be fewer of the heavier shape and more of the lighter shape in order for the two strings of the mobile to balance. But it is *not* always the case that the number of squares is the same as the weight of the triangles, as it was in the first case we examined. This initial experience with considering the balance of these mobiles will help students in making sense of the last line of the *Thinking Out Loud* dialogue.

Have students read the dialogue silently and ask three volunteers to act out the dialogue for the class. At the end of the dialogue, students should consider the prompt in the *Pausing to Think* box to see how the three equations, $\frac{g}{h} = \frac{5}{2}$, $\frac{h}{g} = \frac{2}{5}$, and $5h = 2g$, all describe the same relationship between h and g.

Lesson at a Glance

Mental Mathematics (5 min)

Launch: Mobiles and *Thinking Out Loud* Dialogue (10 min)

· Students build simple mobiles relating pairs of values.

· Students read and act out a *Thinking Out Loud* dialogue about writing an equation for a proportional relationship.

Student Problem Solving and Discussion (30 min)

· Give students time to work through the Important Stuff and explore additional problems.

· Discuss ways of writing equivalent equations for proportional relationships.

Unit 8 Related Activity: Making MysteryGrids (See page T37 and Student Worktext page 50.)

A common mistake when writing proportional relationships is for students to read the relationship "2 hours for 5 gardens" and write the equation $2h = 5g$. This is a common mistake because it seems to make linguistic sense read as 2 hours "equals" 5 gardens. But $2h$ does not mean "2 hours." The variable h stands for the number of hours, so $h = 2$. As Lena points out in the dialogue, $2h$ means the number of hours is doubled.

There are many other ways to express this proportional relationship, such as $\frac{g}{5} = \frac{h}{2}$ and $\frac{2}{h} = \frac{5}{g}$. These forms of the equation are not included in the *Thinking Out Loud* dialogue because the characters are demonstrating their use of the repeating and generalizing strategy. However, if students generate equations of this form, use the opportunity to show that these equations are equivalent to the rest.

Algebraic Habits of Mind

Describing Repeated Reasoning

Repeating and generalizing is an effective strategy for figuring out what algebraic equation to create for a problem situation. Perhaps more importantly, using numerical examples helps students enter into and make sense of a problem.

Whenever a problem asks students to find the relationship between two covarying quantities, the first thing to do is determine the *nature* of the relationship. Working with numerical examples helps students understand the idea of proportions. For example, in "Hiroshi can read 2 books in 7 days," students must understand that 2-in-7 means that reading another 2 books takes another 7 days. Likewise, if a student encounters a problem statement such as "Raj is 4 years old and Mali is 6 years old," numerical experimentation will show that their ages are *not* proportional (when Raj is 8 years old, Mali won't be 12 years old). Whenever students write an equation showing a proportional relationship, make sure they have gone through the process of repetition by considering several sets of equivalent fractions before they make their generalization.

For example, in order to validate Lena's equation, $\frac{h}{g} = \frac{2}{5}$, students might identify and reason through equivalent fractions to show that this equation describes the relationship. To validate Jay's equation, $5h = 2g$, students may propose scaling $\frac{g}{h} = \frac{5}{2}$ by multiplying both sides by 2 (giving $\frac{2g}{h} = 5$) and then multiplying both sides by h to get $2g = 5h$.

Student Problem Solving and Discussion

PROBLEMS 10, 13, & 14 give students scenarios in which values are related proportionally. Whenever students are asked to write an equation for the relationship, encourage them to use the repeating and generalizing strategy by writing some equivalent fractions before expressing the general pattern with variables. At the end of class, have students share the different equations they wrote for each problem and discuss the equivalence of the equations.

These questions ask students to think about equivalent ways to represent a proportional relationship:

» **In problem 10, how can you tell that the relationship between white and pink carnations is proportional (and therefore can be represented as a fraction)?** It makes sense that multiplying the number of white carnations by some scale factor (a number) means the number of pink carnations should also be multiplied by the same scale factor (number) in order to make the bouquet. So, for example, a bouquet might contain 8 white and 6 pink carnations. Present students with an incorrect example: "How do you know that a bouquet won't contain 10 white carnations and 9 pink carnations?" (This example intentionally has 1 more white carnation than pink carnations to mimic the 4 white and 3 pink carnations in the problem.) Listen for students who reason that this example doesn't work because there are 3 times as many pink carnations, so there should be 3 times as many white carnations as well.

» **For the equations in problems 10, 13, and 14, what are some other equivalent ways of expressing each relationship?** This question reinforces the idea that proportional relationships can be thought of structurally as equivalent equations. For example, in problem 13, students can start with the information ("Hiroshi can read 2 books in 7 days") and build equivalent fractions such as $\frac{2}{7} = \frac{4}{14} = \frac{20}{70}$ and recognize that all of these fractions are of the form $\frac{b}{d}$. There are infinitely many ways of writing the relationship between the two variables in each case, including $\frac{b}{d} = \frac{4}{14}$, $70b = 20d$, and so on.

» **Make up other scenarios of two quantities with a proportional relationship.**

» **Make up scenarios of two quantities that do *not* have a proportional relationship.** Some examples include your age and the age of a younger sibling, the number of years you own a car and its value, or height and age.

Lesson 8:
Fractions and Graphs (Rates of Change)

PURPOSE

In this lesson, students learn to translate between two ways of representing proportional relationships: graphs and equations. Proportional relationships are linear relationships that include (0, 0). Students encounter several central ideas about the slope of a line in graphs of such relationships. For example, if such a line passes through the point (2, 3), then it will also pass through the points (4, 6), (6, 9), and (8, 12). There is a proportional relationship between the x-coordinate and the y-coordinate. In this example, one might write that $\frac{x}{y} = \frac{2}{3}$ or $\frac{y}{x} = \frac{3}{2}$. Later, students will formalize the second of these fractions, $\frac{3}{2}$, as the slope of the line that passes through (0, 0) and (2, 3). This slope is the rate of change of the solution points that make up the graph and is a way of describing the steepness of the line. The proportional relationship is also a way of writing the equation for the graph of the line. This equation may be written as $\frac{y}{x} = \frac{3}{2}$, $y = \frac{3}{2}x$, $3x = 2y$, or any number of other ways; all of these equivalent forms describe the relationship between the x- and y-coordinates of the points on the graph. This lesson foreshadows students' study of slope of linear equations in Unit 9: *Points, Slopes, and Lines.*

 **Mental Mathematics** *Begin each day with five minutes of Mental Mathematics (pages T51–T64). Working with factors helps prepare students for later units, especially Unit 10: Area Model Factoring.*

Launch: Comparing and Calculating with Rates

Display "Comparing Rates" (page T42). Show the top half of the page, Rates Displayed in a Graph, and have the class discuss the questions. Listen for students who explain, for example, that Jacob must be traveling slower because in the same amount of time, he covers less distance. Encourage students to identify specific points on the graph as evidence. For instance, at the 2-hour mark, Jacob has traveled 10 miles while Mali has traveled close to 50 miles.

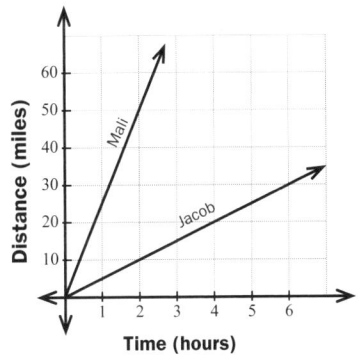

Lesson at a Glance

Preparation: Prepare to display "Comparing Rates" on page T42.

Mental Mathematics (5 min)

Launch: Comparing and Calculating with Rates (15 min)

· Students compare rates on a graph and use information about a rate to fill out a table.

Student Problem Solving and Discussion (25 min)

· Give students time to work through the Important Stuff and explore additional problems.

· Discuss the connection between proportions and graphs.

Unit 8 Related Activity: Making MysteryGrids (See page T37 and Student Worktext page 50.)

Then show Rates Displayed in a Table and fill out the table together (answers shown here). Use this opportunity to make sure students understand the phrase "miles per hour" as the number of miles Mali travels in one hour. The graph on the top half of the page corresponds to these figures, so students can also look for how the graph supports the values found in the table. Students can also discuss how the values in the table show that hours and miles have a proportional relationship. Students can build equivalent fractions, such as $\frac{48}{2} = \frac{96}{4} = \frac{24}{1}$. Generalizing, students can write the equation $\frac{miles}{hours} = 24$ and connect this fraction to the fact that the rate at which Mali is traveling is 24 miles per hour. If there is time, create and analyze a similar table for Jacob to find his rate of travel (5 mph).

Hours	Miles
2	48
4	96
$\frac{1}{2}$	12
$\frac{1}{4}$	6
1	24

Student Problem Solving and Discussion

PROBLEMS 1 & 2 provide graphs for students to use to answer questions about proportional relationships. In PROBLEM 3, students must draw their own graph. As students solve these problems, listen for ways they consider the graphs using proportional reasoning. For example, in drawing the graph for problem 3, a student might notice that "Since a barista makes 8 coffee drinks every 10 minutes, that must mean she makes 4 drinks every 5 minutes, so (10, 8) and (5, 4) must both be on the graph."

These questions ask students to think about connections between proportions and graphs:

> » **Miles per hour, beats per minute, and dollars per hour are all rates. What are some other rates you might use to measure?** If there is time, you may wish to create tables, equations, and/or graphs for specific rate values of students' choosing.

> » **The way you say a rate (like "miles per hour") provides a clue for how to set up a proportion. Use this strategy to write equations for problems 1, 2, and 3. Do the equations match other equations you wrote for these relationships?** For example, problem 1b asks students about how many dollars per hour Imani makes. Using this clue, a student might set up the proportion $\frac{dollars}{hour} = \frac{30}{4}$.

Student Reflections & Unit Assessment

Before the Unit Assessment, have students reflect on their learning:

What are some things you learned in this unit?

What questions do you still have?

Reflections can be done orally, on paper, or some combination of both. Use feedback from students to help them identify the big ideas from the unit and to select Additional Practice problems to help them prepare for the Unit Assessment included on pages T47 & T48. Before giving this assessment, consider spending a class period working through the Unit Additional Practice problems.

Since the Snapshot Check-in, students have:

- Solved equations with fractions by scaling both sides of an equation.
- Applied their understanding of solving equations with fractions to make sense of problems with fraction addition.
- Written and solved problems about proportional relationships.
- Interpreted graphs representing proportional relationships.

Throughout Unit 8, students have focused on developing the following Algebraic Habits of Mind:

- **Using Tools Strategically**—Students have used the number line as a visual model to judge the reasonableness of a solution and area models to represent fraction multiplication, providing a way to make sense of multiplying fractions by multiplying numerators and multiplying denominators, and to organize calculations.
- **Seeking and Using Structure**—Students have considered equivalent forms of rational expressions and extended their understanding of maintaining the balance of an equation to solving equations with fractions. They have also used equivalent fractions to help write equations for proportional relationships and to read rate information from graphs.
- **Puzzling and Persevering**—Students have encountered many familiar puzzle types that now include fractions. Students have also judged the reasonableness of solutions, approached new problems by connecting them to familiar ones, and found entry points to problems through numerical examples.
- **Describing Repeated Reasoning**—Students used the strategy of repeating and generalizing to give themselves enough concrete experience with a proportional relationship to recognize its regularity and write a general equation for it.

Preparation

- (*optional*) "The 60th Triangular Number" is provided on page T43 to use during classroom discussion.

- (*optional*) "10 × 10 Squares" is provided on page T44 to use as a display during classroom discussion or to photocopy and provide to students to use as they explore problem 8.

Mental Mathematics (5 min)

Student Exploration and Discussion
(40 min)

- Provide time for students to explore problems 1–7.

- Ask students to share observations about the connection between triangular and square numbers.

- Provide time for students to explore problems 8–13 and discover a way to find the value of any triangular number.

- Share and discuss formulas that students discover.

Unit 8 Related Activity: Making MysteryGrids (See page T37 and Student Worktext page 50.)

→ EXPLORATION
Triangular Numbers

PURPOSE

This Exploration culminates a series of Explorations in which the sequence of numbers 1, 3, 6, 10, 15, 21, ... has occurred in a variety of contexts including Color Towers 1 (Unit 1), Coin Combinations (Unit 3), Toothpick Rows 2 (Unit 3), Street Paths (Unit 6), and Staircase Patterns (Unit 7). Students have been repeatedly exposed to this sequence so that they may start to recognize it and wonder about how it can show up in such varied contexts. This Exploration formally names this sequence the "triangular numbers," and students finally discover a formula for finding the nth triangular number by experimenting and describing repeated reasoning. As students become familiar with special sets of numbers (prime numbers, odd numbers, square numbers, and so on), they are better able to make connections and find patterns in numbers.

Mental Mathematics *Begin each day with five minutes of Mental Mathematics (pages T51–T64). Even on an Exploration day, Mental Mathematics is an important way to gear students up to engage in good mathematical thinking.*

Student Exploration and Discussion

There are many ways to find an equation for the nth triangular number, T_n. In the Unit 7 Exploration Staircase Patterns, students have already seen one way. In this Exploration, they learn another way to find T_n by observing that the sum of any two consecutive triangular numbers is a square number.

After students have had time to work on **PROBLEMS 1–7**, gather the class for a brief discussion so that students have a chance to share their thoughts on why it makes sense that two consecutive triangular numbers sum to a square number. Discuss the connection between triangular and square numbers.

» **Is it just a coincidence that the square numbers show up in this way, or is there a connection you can explain?** Students can show geometrically how two consecutive triangles can be oriented to form a square. An example that connects to problem 5 is shown at right.

» **Two consecutive triangles aren't the same size; how is it that they make a square whose sides are all the same?** Make sure students see that each triangle is *not* half the square. The base and height of the smaller triangle are smaller by 1 unit; the larger triangle makes up for this "missing" unit at each end of its hypotenuse.

» **How are the sizes of any two consecutive triangular numbers connected to the size of the resulting square?** This idea and the associated notation is the focus of problem 7. Students should see that, for example, the 7th triangle's legs are each 7 units long and the 8th triangle's legs are 8 units long; these triangles will form a square whose sides are all 8 units long, the same length as the legs of the larger triangle.

» **Estimate T_{60}.** Use "The 60th Triangular Number" on page T43 to give students a visual reference. Ask students to estimate T_{60} by saying a number that they know is too high and one that is too low, then justifying those.

Finding a formula for T_n

In **PROBLEMS 8–11**, students use the fact that the sum of any two consecutive triangular numbers is a square number to find and express a way to find any triangular number, T_n. There are many possible solution pathways for finding a formula for T_n. Here are just four of them. In each case, the problem is first worked out for a specific case—finding T_9 and T_{10} from knowing their sum is 100—as presented in problem 8. Then, each example shows how this thinking generalizes to finding T_n.

Method 1 (presented in the Answer Key):
The figures T_9 and T_{10} together form a 10×10 square. They are each about half of the square except that the T_{10} figure includes the extra circles along the diagonal. There are 10 circles along the diagonal. So, T_9 is half of $(100 - 10)$, half of $(n^2 - n)$. And T_{10} is 10 more than that. That is, T_n is n more than $\frac{n^2-n}{2}$. Stated algebraically, $T_n = \frac{n^2-n}{2} + n$.

Method 2:
The figures T_9 and T_{10} together form a 10×10 square. They are each about half of the square except that the T_{10} figure includes all (not half) of the circles along the diagonal. So T_{10} is half of 100 plus 5 extra half-circles. Stated algebraically, $T_n = \frac{n^2}{2} + \frac{n}{2}$.

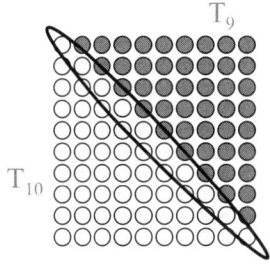

Method 3:
The figures T_9 and T_{10} together form a 10×10 square. They are each about half of the square except that the T_{10} figure includes an extra column of circles on the left. T_{10} is half of $(100 - 10)$ plus 10 extra circles. Stated algebraically, $T_n = \frac{n^2-n}{2} + n$.

Method 4:
We know two relationships: $T_9 + T_{10} = 100$ and $T_9 + 10 = T_{10}$. Using substitution, we get $T_9 + T_9 + 10 = 100$. Solving for T_9, we see $T_9 = 45$, so $T_{10} = 45 + 10 = 55$. In general, $T_{n-1} + T_{n-1} + n = n^2$. So $T_{n-1} = \frac{n^2-n}{2}$. And because T_n is n more than that, $T_n = \frac{n^2-n}{2} + n$.

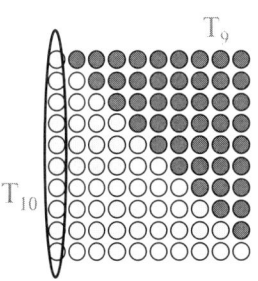

Describing Repeated Reasoning

The crux of this Exploration lies in problems 8–11. Students have just discovered that the sum of any pair of consecutive triangular numbers is a square number, and they must experiment to figure out how to start with a square number to find its two component triangular numbers. In problem 8, students find T_{10} (the 10th triangular number) by knowing that $T_9 + T_{10} = 10^2$. This requires several steps of reasoning with several pathways students might use to arrive at a solution. Students then recreate this reasoning to find T_3 and T_{20} before expressing their process as an algebraic expression for T_n.

Students who exhibit this habit of mind understand that algebraic expressions and formulas are "pattern describers." They understand how to separate the details of a specific case from a general pattern and understand that the solutions for *specific* cases are just numbers, but the *process* of arriving at those solutions is more powerful because it is a solution for a *general* case, for *all* numbers (in the domain). Students who are proficient look for similarities between different cases and test their reasoning for smaller and larger cases. They understand the connection between the formula they have found and the problem at hand; they know what the formula is for and how and when to use it.

Listen for strategies students use to break apart a square into its two component triangles. Use "10 × 10 Squares" on page T44 to facilitate discussion. Help students describe their strategies precisely enough so that the variety of solution pathways generated by the class is evident. Provide opportunities for students to talk about connections they see among the solutions and share their observations and insights.

» **What similarities do you notice in your classmates' strategies?** This depends on what students discovered and presented. Even if only two strategies were presented, students may compare the ways in which the strategies used pictures or expressions or numbers. Or they may have recognized that of the two consecutive triangular numbers whose sum is n^2, one had to be less than $\frac{n^2}{2}$ and the other had to be greater. Even if not all students came up with a complete strategy for finding T_n, they might compare their work to the other strategies presented and consider the equivalence of the final algebraic results.

» **How did finding T_{10} and T_{20} help you find an expression for T_n?** Give students the opportunity to explain their process for finding T_{10} and T_{20}, and listen for students who explain how they connected this thinking to finding T_n. Listen for the idea that n was used in place 10 and 20 and that finding an expression for T_n meant figuring out how all the other numbers used depended on 10 and 20 to come up with a general formula.

» **Find T_{60}.** Use "The 60th Triangular Number" on page T43 as a visual reference. Listen for students who are able to find T_{60} by reapplying the process they used to find T_{10}, T_{20}, and T_n. Listen also for students who understand they can use their formula for T_n to find this value: $T_{60} = 1830$. Compare this result to the estimations made earlier.

» **In the Unit 7 Exploration Staircase Patterns, you examined the area of the rectangle formed by putting two triangles of the *same* size together to get a rectangle with length n and width $n + 1$. Doing this, you found another formula for T_n (although it didn't have that label): $T_n = \frac{n(n+1)}{2}$. Compare this with the formula you found today.** The expressions are equivalent.

PROBLEM 12 presents a word problem (about the sum of consecutive counting numbers) whose structure is meant to invoke the use of triangular numbers.

PROBLEM 13 asks students to invent their own situation for which they might want to know the nth triangular number. For example, "If I run 1 mile on the first day of training, 2 miles on the second day, 3 miles on the third day, and so on, then how many total miles will I have run by the end of 31 days?" is an example of a problem in which the answer is T_{31}. A different kind of example to consider with students is "How many 2-topping pizza combinations are possible from a list of 6 toppings?" The answer is T_5. This problem's structure is similar to the problem explored in the Unit 1 Exploration Color Towers, in which students found all the ways a tower of 2 blue and 4 white blocks could be arranged. In the case of the pizza toppings, if the 6 toppings are A, B, C, D, E, and F, there are 5 different pizzas that can be made using topping A (AB, AC, AD, AE, and AF), 4 using B (not counting BA, since that combination has already been counted, so BC, BD, BE, and BF), 3 using C, and so on. Other examples are found in the Answer Key.

You may want to use the following prompt to help students see how problem 12 connects with triangular numbers:

» **How is the story in problem 12 related to the shape of a triangle?** Students can translate this problem directly to the process of finding the sum of consecutive numbers. Rows of the triangle represent days in the story. Finding the number of rocks in the collection after the 10th day is the same as finding T_{10}.

Further Exploration

PROBLEMS 14–17 extend the Exploration to pentagonal numbers (problem 14 asks students to invent a descriptive name). Then **PROBLEM 18** asks students to invent and describe their own set of numbers. This can be a very creative mathematical activity. Students don't need to limit themselves to sequences for which they can write a formula. Encourage them to explore to find a pattern they find interesting (they can name it after themselves), and ask them to describe what makes the pattern interesting. Students may explore problem 18 without completing problems 14–17.

Preparation: *(optional)* It might help
students to have graph paper to draw
squares.

Mental Mathematics (5 min)

Launch: Introducing Squares of Squares
(5 min)

· Pose and explore the main question of the
Exploration: Into how many squares can a
square be cut?

Student Exploration and Discussion
(50 min)

· Provide time for students to explore
problems 1–5.

· Discuss student responses to problems
3–5 in which students build an argument
for being able to cut a square into any
number of squares greater than 7.

Unit 8 Related Activity: Making
MysteryGrids (See page T37 and Student
Worktext page 50.)

→ EXPLORATION
A Square of Squares

PURPOSE

In this Exploration, students dissect a square into smaller squares
and explore the question: Into how many squares can a square be
cut? Students work with patterns of even and odd numbers and build
an argument that extends to cover infinitely many cases. Taking a
complicated problem and organizing it into simpler subcases is a useful
problem-solving strategy and develops the habit of mind of seeking and
using structure in a problem. In the Further Exploration, students apply
this same strategy to an analogous problem with equilateral triangles.

Mental Mathematics *Begin each day with five minutes of Mental
Mathematics (pages T51–T64). Mental Mathematics highlights many aspects of
algebraic thinking, including identifying patterns, exploring properties of operations,
and exploring multiple ways of expressing equivalent quantities.*

Launch: Introducing Squares of Squares

Before students open their Worktexts, pose the problem **Suppose I start with a
square and I cut it so that every piece is a smaller square. Can I cut a square into
two squares? Into how many squares *can* a square be cut?** Invite students to draw
ideas on the board.

Students will likely suggest the square numbers and might
think the problem is finished. Wait silently for someone to
suggest a different idea. The ideas don't have to be the ones in
the Student Worktext. For example, a student might draw a
picture like the one shown to the right.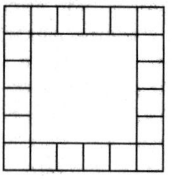

If, after a long time, the only examples are still perfect squares, give the
small hint that the squares don't have to all be the same size. Just as students
start to generate other ideas, instruct them to open their Worktexts.

Student Exploration and Discussion

In **PROBLEM 3**, students examine Pattern 1 (from problem 2) and find that the
pattern shows that any even number of squares greater than or equal to 4 can
be cut from a larger square.

Pattern 1:

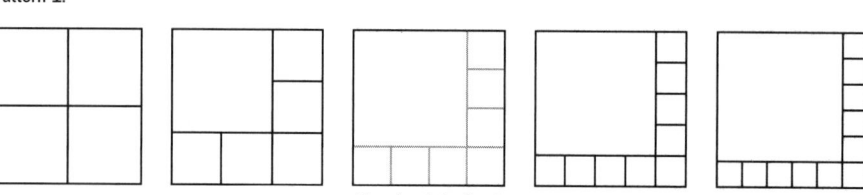

(draw the missing figure)

We know that *every* even number greater than or equal to 4 will be covered by Pattern 1 because the pattern adds two squares each time it grows. Students can use similar reasoning for **PROBLEM 4**, in which they find that Pattern 2 shows that any odd number of squares greater than or equal to 9 can be cut from a larger square.

Pattern 2:

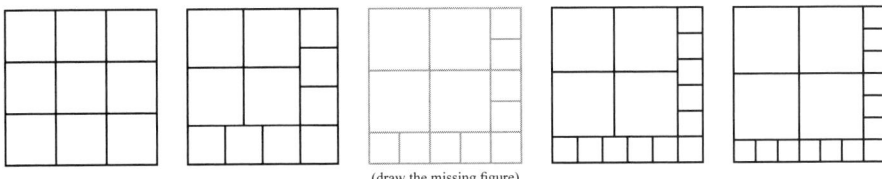

(draw the missing figure)

Students use these problems to explain that Patterns 1 and 2 are enough to establish that any number of squares other than 2, 3, 5, and 7 can be cut from a larger square. Students who go on to the Further Exploration will see that it is also possible to cut a square into 7 squares.

These questions help students explore more ideas about even and odd numbers:

» **In the patterns in problem 2, the total number of smaller squares along the bottom and side of the original square must be odd. Why must the number of smaller squares be odd?** There is one small square in the bottom right corner. The number of remaining squares along the bottom must equal the number of remaining squares along the side of the square of larger squares, because all sides of a square are equal. This means the total number of smaller squares is a number one more than an even number, so it must be odd. This idea was also explored in the Unit 4 Exploration Building Squares.

» **How could you cut a square into 200 squares?** Since 200 is even, it can be built using Pattern 1 with one large square and 199 small squares. The small squares would be arranged with 1 corner square, 99 more along the bottom, and 99 more on the side. This is not the only way to cut a square into 200 squares. In general, a square can be cut into any combination of a perfect square + an odd number. So, for example, the 200 squares could start with 31 squares along two of the edges (1 in the corner, 15 on the bottom, 15 on the side), and the remaining part could be cut into 169 squares (that's 13^2). Since the number of squares along the two edges must always be odd, the perfect square must be odd because the sum of two odd numbers is even. Other combinations are also possible.

» **How could you cut a square into 99 squares?** 99 is odd, so it can be built using Pattern 2 with 4 larger squares and 95 smaller squares (1 in the corner, 47 on the bottom, 47 on the side). Again, this is not the only way to cut a square into 99 squares. The square can be cut into any combination of an *even* perfect square + an odd number. Other combinations are also possible.

Algebraic Habits of Mind

Seeking and Using Structure During this Exploration, students build an argument, describing just two patterns, to show that they can cut a square into any number of even squares greater than or equal to 4 and any number of odd squares greater than or equal to 7. Students use the given patterns to make a generalization; first they notice that they can cut a square into *specific* numbers of squares; later they extend the pattern, establishing an argument that covers infinitely many cases. In order to make the claim for an infinite number of cases, students must take care to show that they know *for sure* that the patterns in their specific cases will extend and cover every larger case.

Further Exploration

In **PROBLEM 6**, students find that a square can be cut into 7 smaller squares. Students may have already found this in their initial exploration in problem 1.

In **PROBLEMS 7 & 8**, students explore an analogous problem about cutting equilateral triangles into smaller equilateral triangles.

Here's an equilateral triangle.   And here are some ways to cut it into smaller equilateral triangles.

   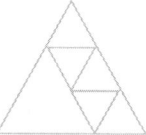   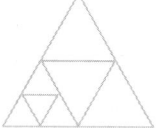   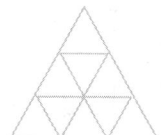

Interestingly, even though the problem uses triangles, the solution can be constructed in a way that is identical to the solution in the case of squares. Problem 8 highlights this similarity by asking students to explore the fact that when an equilateral triangle is cut into congruent equilateral triangles, the number of smaller triangles is always a perfect square. As it mentions in the Answer Key, students may recall examining related ideas in the Toothpick Rows 2 Exploration (from Unit 3) and in Building Squares (from Unit 4).

Activity:
Making MysteryGrids

PURPOSE

When students *make* a puzzle, they must think more deeply about the structure, the nature of the clues, what makes a solution unique, and the strategies they can use to solve the puzzle, even with harder clues. The goal of this activity is to foster such deeper analysis.

Activity at a Glance

Timing: This activity is intended to be done in short 5- to 10-minute sessions throughout Unit 8.

Preparation: Provide students with graph paper or copies of "Blank MysteryGrids" (available in the Resources PDF and on page T38) so they can make their own MysteryGrids. You may want to provide markers or colored pencils so that students can outline the cages in their puzzles clearly.

Activity Suggestions

Activity instructions are on page 50 of the Student Worktext. The basic idea is the same as for making up any kind of puzzle. First, students make up the solution, which is like making a Latin Square puzzle. Then students mark off cages and make up clues for them. Finally, students try to solve their own puzzle themselves, to test it to see if their puzzle has a unique solution before giving it to others to enjoy.

Encourage students to create MysteryGrid puzzles that are hard enough to be fun, but easy enough that they believe their peers can solve them.

Some ways to make a MysteryGrid harder are to start with a larger grid (say 5×5 or larger), have larger cages, or choose more challenging clues (for example, multiplication clues are often much easier than subtraction clues).

An important part of making a puzzle is making sure it has a unique solution. Encourage students to find their own ways of adjusting their cages and clues until their puzzle has a unique solution. One simple way to adjust a puzzle is to use more single-cell cages, but that, of course, also makes it easier. Some puzzles might only require small adjustments to the cages to make a puzzle with a unique solution; other puzzles will need a fresh start. As students work, listen for those who develop approaches for making puzzles with unique solutions. For example, students will find that making cages such as the pair of highlighted cages at the right will lead to a MysteryGrid with a non-unique solution, because the elements 2 and 4 in those two columns will be interchangeable.

MysteryGrid **1, 2, 3, 4**

1	2	4	3
4	1	3	2
2	3	1	4
3	4	2	1

Variations

- Students can make larger MysteryGrids: 5×5, 6×6, or even larger.
- Students can use positive or negative integers, decimals, or fractions as elements and clues.
- Students can even use algebraic expressions in their puzzles, as they saw in some Unit 7 MysteryGrids (with elements such as 1, x, and $2x$). Students will see more MysteryGrids with algebra in later units.

Blank MysteryGrids (Activity: Making MysteryGrids)

3×3

MysteryGrid ____, ____, ____

MysteryGrid ____, ____, ____

MysteryGrid ____, ____, ____

 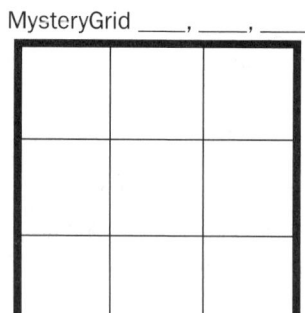 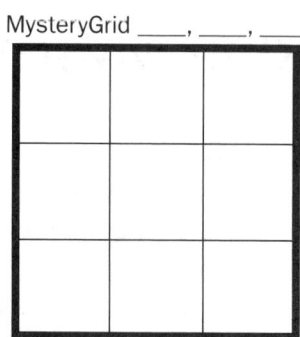

4×4

MysteryGrid ____, ____, ____, ____

MysteryGrid ____, ____, ____, ____

MysteryGrid ____, ____, ____, ____

5×5

MysteryGrid ____, ____, ____, ____, ____

MysteryGrid ____, ____, ____, ____, ____

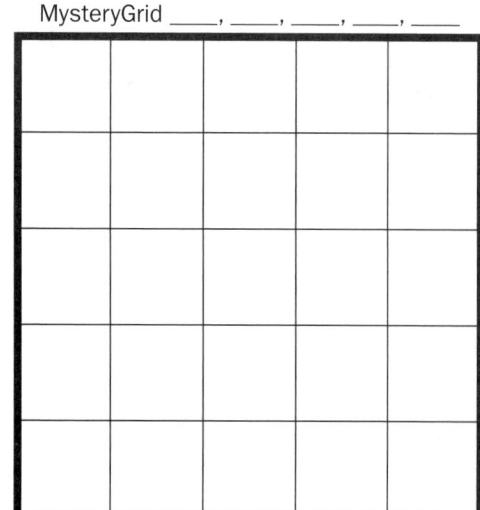

Cuisenaire Tables (Lesson 1)

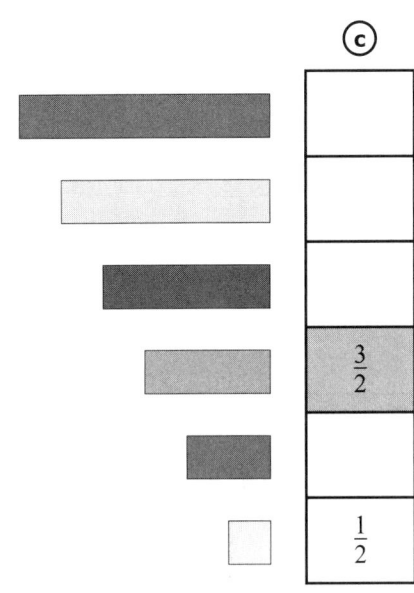

	ⓒ
	$\frac{3}{2}$
	$\frac{1}{2}$

① These rods can be arranged into a square, as shown on the left.
In each column, use the given information to find values for the lengths of the rest of the rods.

As you can see from comparing the rods' lengths, if a red rod's length is 10, then the white rod must be 5 and the light green rod must be ____.

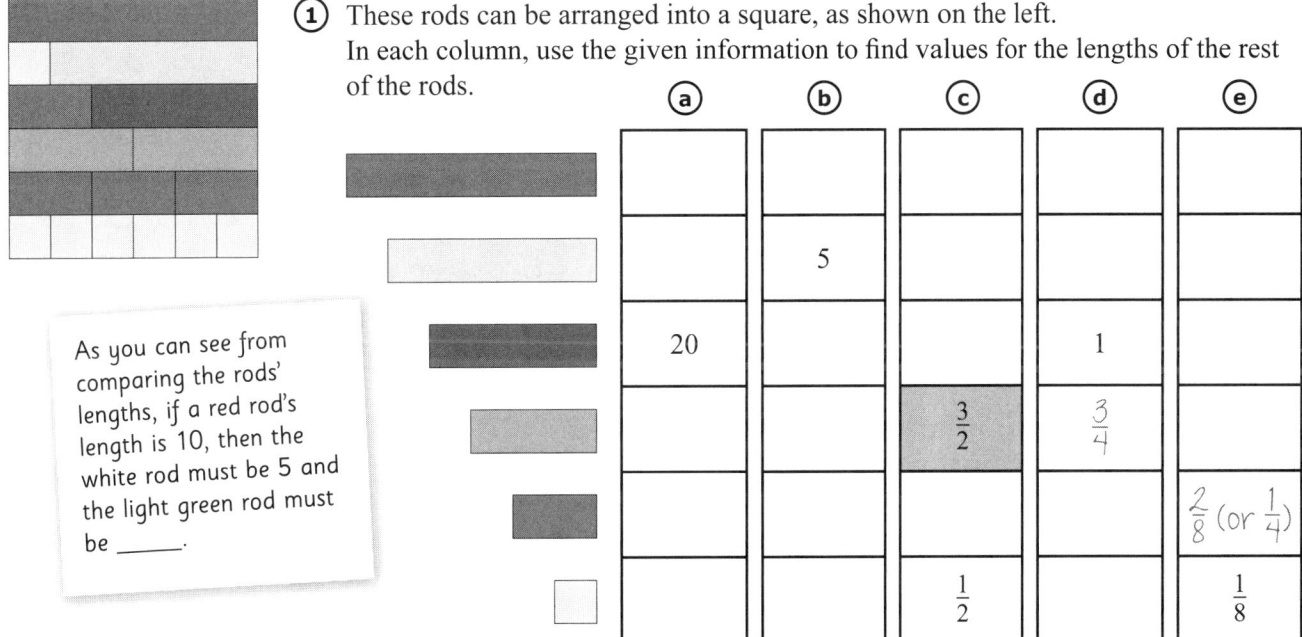

	ⓐ	ⓑ	ⓒ	ⓓ	ⓔ
		5			
	20			1	
			$\frac{3}{2}$	$\frac{3}{4}$	
					$\frac{2}{8}$ (or $\frac{1}{4}$)
			$\frac{1}{2}$		$\frac{1}{8}$

	a	b	c	d	e
		12			$\frac{6}{5}$
					$\frac{9}{10}$
			3		
	1			$\frac{3}{5}$	

How do these models show the quantity $\frac{4}{3}$?

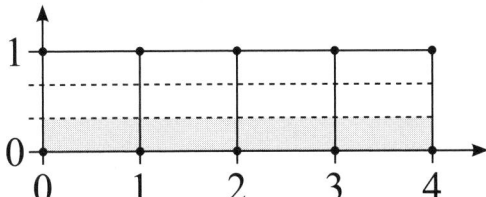 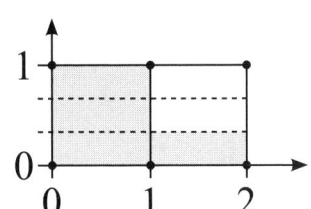 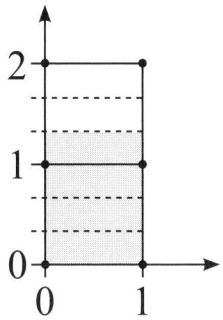

Some models that show the quantity $\frac{1}{2} \cdot \frac{4}{3}$:

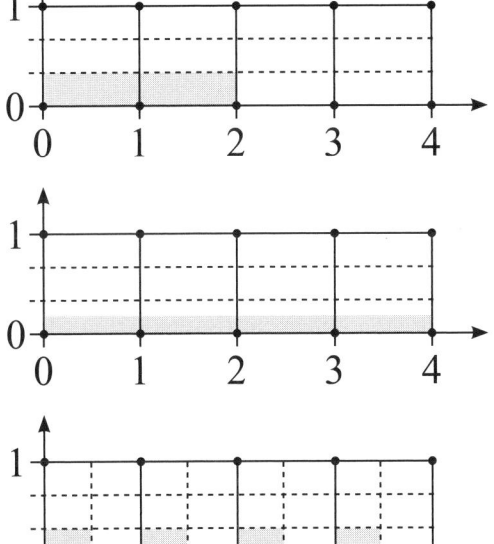 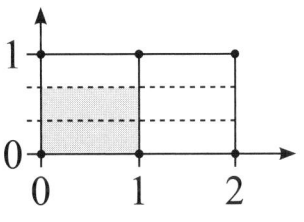 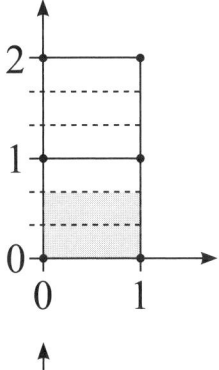

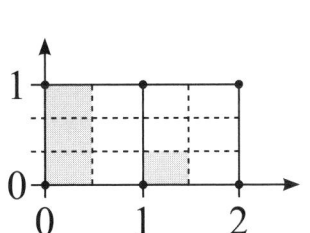 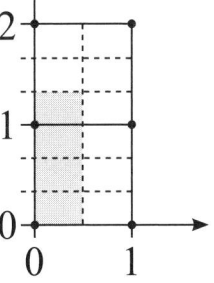

Rates Displayed in a Graph

Mali and Jacob are both traveling without using fossil fuels. This graph shows each of their distances over time.

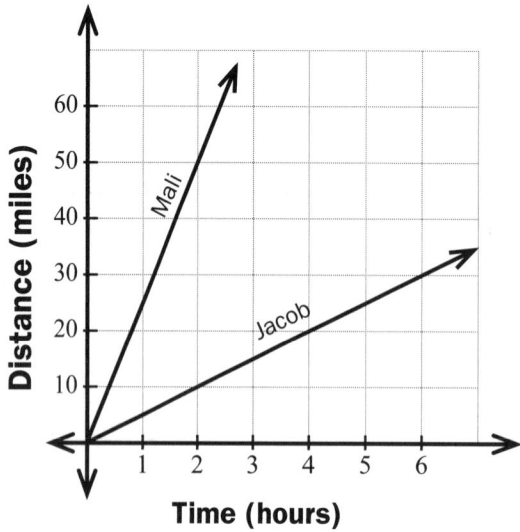

One of them is riding a bicycle and the other is jogging.

What is Mali doing? How did you decide?

Rates Displayed in a Table

Mali travels 24 miles per hour.

Complete this table to show how far she can go in various lengths of time.

Hours	Miles
2	
4	
$\frac{1}{2}$	
$\frac{1}{4}$	
1	

The 60th Triangular Number (Exploration: Triangular Numbers)

Name: _____

(1) Circle **all** the expressions that show $\frac{5}{3} + \frac{5}{3} + \frac{5}{3}$.

(A) $\frac{5}{3} \cdot \frac{3}{3}$

(B) $3 \cdot \frac{5}{3}$

(C) $\frac{5 + 5 + 5}{3 + 3 + 3}$

(D) $\frac{5 + 5 + 5}{3}$

(2) Circle **all** the expressions that show $\frac{6}{7} \cdot 2$.

(A) $\frac{6}{7} + \frac{6}{7}$

(B) $6 \div 7 \cdot 2$

(C) $\frac{6 \cdot 2}{7 \cdot 2}$

(D) $6 \cdot 2 \div 7$

(3) Multiply each number by 3.

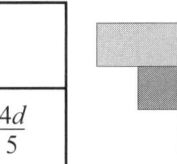

	$\frac{1}{3}$	$\frac{7}{3}$	$\frac{277}{3}$	$\frac{a}{3}$	$\frac{85b}{3}$	$\frac{1}{5}$	$\frac{2}{5}$	$\frac{c}{5}$	$\frac{4d}{5}$

(4) Write the equation for the problem that this area model represents.

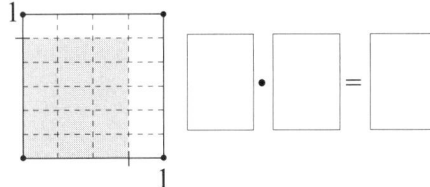

(5) Draw a rectangle whose area is $\frac{2}{3} \cdot \frac{1}{3}$. Then find the product.

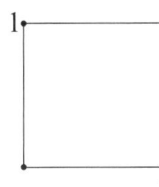

(6) MysteryGrid **$\frac{1}{3}$, 1, 3**

3	$\frac{1}{3}, \bullet$	
$\frac{2}{3}, -$		4, +
1, $\bullet$		

(7) Who Am I?
- I am less than $\frac{1}{2}$.
- My numerator and denominator are each one digit.
- The product of my numerator and denominator is 10.

$$\frac{numerator}{denominator} = \frac{\square}{\square}$$

(1) Circle **all** the expressions that show $\frac{5}{3} + \frac{5}{3} + \frac{5}{3}$.

(A) $\frac{5}{3} \cdot \frac{3}{3}$

(B) $3 \cdot \frac{5}{3}$ ⬭

(C) $\frac{5+5+5}{3+3+3}$

(D) $\frac{5+5+5}{3}$ ⬭

(2) Circle **all** the expressions that show $\frac{6}{7} \cdot 2$.

(A) $\frac{6}{7} + \frac{6}{7}$ ⬭

(B) $6 \div 7 \cdot 2$ ⬭

(C) $\frac{6 \cdot 2}{7 \cdot 2}$

(D) $6 \cdot 2 \div 7$ ⬭

(3) Multiply each number by 3.

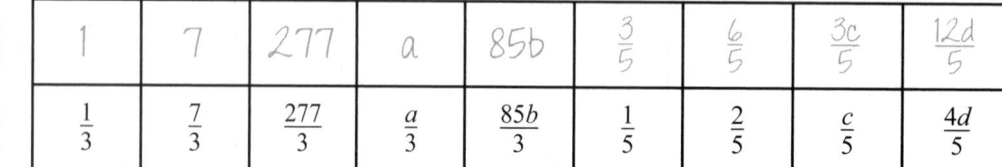

1	7	277	a	$85b$	$\frac{3}{5}$	$\frac{6}{5}$	$\frac{3c}{5}$	$\frac{12d}{5}$
$\frac{1}{3}$	$\frac{7}{3}$	$\frac{277}{3}$	$\frac{a}{3}$	$\frac{85b}{3}$	$\frac{1}{5}$	$\frac{2}{5}$	$\frac{c}{5}$	$\frac{4d}{5}$

(4) Write the equation for the problem that this area model represents.

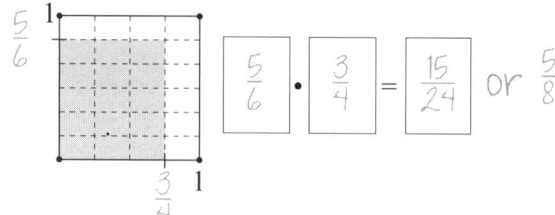

$$\frac{5}{6} \cdot \frac{3}{4} = \frac{15}{24} \text{ or } \frac{5}{8}$$

(5) Draw a rectangle whose area is $\frac{2}{3} \cdot \frac{1}{3}$. Then find the product.

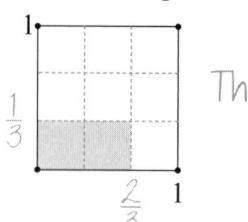

The product is $\frac{2}{9}$.

(6) MysteryGrid $\frac{1}{3}$, **1, 3**

3 — 3	$\frac{1}{3}$, • — 1	$\frac{1}{3}$
$\frac{2}{3}$, − — 1	$\frac{1}{3}$	4, + — 3
1, • — $\frac{1}{3}$	3	1

(7) Who Am I?
- I am less than $\frac{1}{2}$.
- My numerator and denominator are each one digit.
- The product of my numerator and denominator is 10.

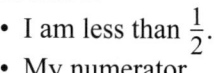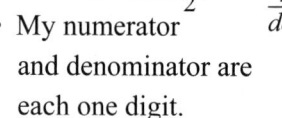

$$\frac{numerator}{denominator} = \frac{2}{5}$$

Unit Assessment

1 These rods can be arranged into a square, as shown on the right. In each column, use the given information to find values for the lengths of the rest of the rods.

	(a)	(b)	(c)	(d)	(e)
		6	3		
	20				
				2	
				$\frac{1}{7}$	

2 Circle the expressions that show one-fourth of 28.

Ⓐ $\frac{28}{4}$ Ⓑ $\frac{20}{4} + \frac{8}{4}$ Ⓒ $\frac{20}{4} + 8$ Ⓓ $\frac{1}{4}(20 + 8)$

Ⓔ $\frac{1}{4} \cdot 28$ Ⓕ $28\frac{1}{4}$ Ⓖ $\frac{8 + 20}{4}$ Ⓗ $\frac{28}{4} \cdot 4$

3 Circle the expressions that are equivalent to $\frac{1}{6}(3x + 24)$.

Ⓐ $\frac{1}{6}(24 + 3x)$ Ⓑ $\frac{1}{6}(27x)$ Ⓒ $\frac{1}{2}x + 4$

Ⓓ $\frac{3x + 24}{6}$ Ⓔ $\frac{24}{6} + \frac{3x}{6}$ Ⓕ $3x + 4$

4 MysteryGrid **$\frac{1}{3}$, $\frac{1}{2}$, 1, 2**

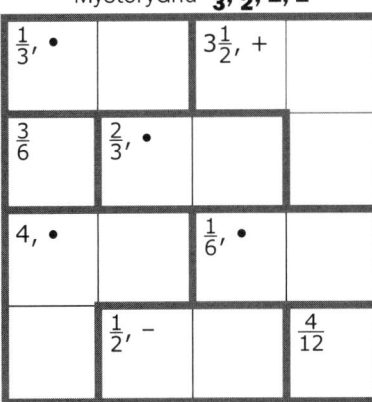

5 Clue: I am $\frac{3}{4} \cdot h$. Where am I?

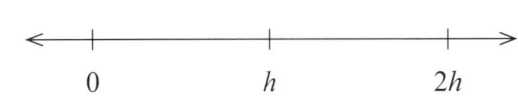

6 Clue: I am $\frac{7}{4} \cdot h$. Where am I?

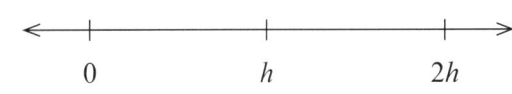

Use the area model to multiply or divide.

(7) $\frac{1}{3}(21w + 3) =$ _____

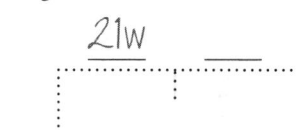

(8) $\frac{20y - 4x + 36}{4} =$ _____

Solve for the variable by scaling or representing the problem on a number line.

(9) $\frac{2}{7}m = 20$

(10) $\frac{3}{5} + \frac{1}{10} = k$

Determine the missing value in each of these equations.

(11) $\frac{22}{16} = \frac{a}{8}$ $a =$

(12) $\frac{35}{b} = \frac{49}{7}$ $b =$

(13) $\frac{c}{8} = \frac{7}{2}$ $c =$

(14) $\frac{9}{27} = \frac{5}{d}$ $d =$

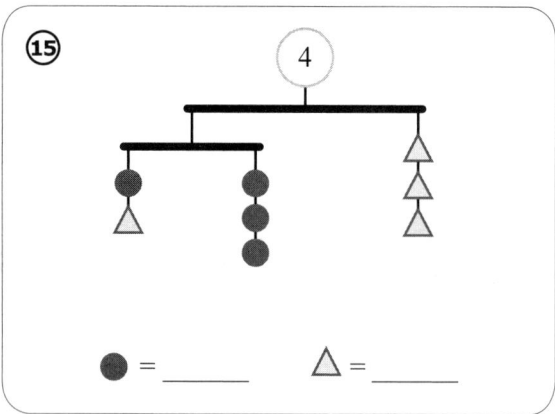

(16) Carla is making a bowl of punch for a birthday party. To make the drink, she mixes a 6-ounce can of frozen pink lemonade and 4 cups of lemon-lime soda.

  (a) Write an equation for the relationship between p, the number of ounces of pink lemonade, and s, the number of cups of soda.

  (b) The store Carla visits only sells frozen pink lemonade in a 9-ounce can. How many cups of soda should she use?

  (c) Carla uses the same recipe to make punch for a school-wide event. She uses a giant can of frozen pink lemonade that contains 20 ounces. How many cups of soda should she use?

1 These rods can be arranged into a square, as shown on the right. In each column, use the given information to find values for the lengths of the rest of the rods.

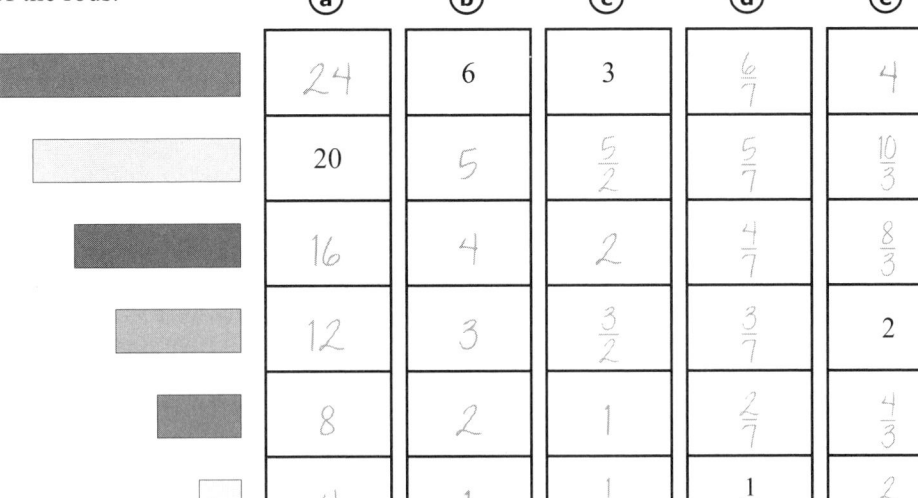

	a	b	c	d	e
	24	6	3	$\frac{6}{7}$	4
	20	5	$\frac{5}{2}$	$\frac{5}{7}$	$\frac{10}{3}$
	16	4	2	$\frac{4}{7}$	$\frac{8}{3}$
	12	3	$\frac{3}{2}$	$\frac{3}{7}$	2
	8	2	1	$\frac{2}{7}$	$\frac{4}{3}$
	4	1	$\frac{1}{2}$	$\frac{1}{7}$	$\frac{2}{3}$

2 Circle the expressions that show one-fourth of 28.

(A) $\frac{28}{4}$ ⟵circled

(B) $\frac{20}{4} + \frac{8}{4}$ ⟵circled

(C) $\frac{20}{4} + 8$

(D) $\frac{1}{4}(20 + 8)$ ⟵circled

(E) $\frac{1}{4} \cdot 28$ ⟵circled

(F) $28\frac{1}{4}$

(G) $\frac{8 + 20}{4}$ ⟵circled

(H) $\frac{28}{4} \cdot 4$

3 Circle the expressions that are equivalent to $\frac{1}{6}(3x + 24)$.

(A) $\frac{1}{6}(24 + 3x)$ ⟵circled

(B) $\frac{1}{6}(27x)$

(C) $\frac{1}{2}x + 4$ ⟵circled

(D) $\frac{3x + 24}{6}$ ⟵circled

(E) $\frac{24}{6} + \frac{3x}{6}$ ⟵circled

(F) $3x + 4$

4 MysteryGrid $\frac{1}{3}$, $\frac{1}{2}$, **1, 2**

$\frac{1}{3}$, • $\frac{1}{3}$	1	$3\frac{1}{2}$, + $\frac{1}{2}$	2
$\frac{3}{6}$ $\frac{1}{2}$	$\frac{2}{3}$, • $\frac{1}{3}$	2	1
4, • 1	2	$\frac{1}{6}$, • $\frac{1}{3}$	$\frac{1}{2}$
2	$\frac{1}{2}$, — $\frac{1}{2}$	1	$\frac{4}{12}$ $\frac{1}{3}$

Where Am I?

5 Clue: I am $\frac{3}{4} \cdot h$. Where am I?

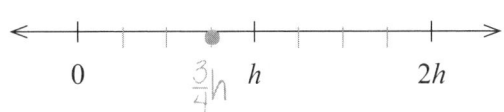

0    $\frac{3}{4}h$ h $2h$

6 Clue: I am $\frac{7}{4} \cdot h$. Where am I?

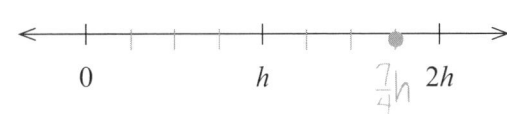

0   h $\frac{7}{4}h$ $2h$

Use the area model to multiply or divide.

⑦ $\frac{1}{3}(21w + 3) =$ ___7w + 1___

	21w	3
$\frac{1}{3}$	7w	1

⑧ $\frac{20y - 4x + 36}{4} =$ ___5y − x + 9___

	5y	−x	9
4	20y	−4x	36

Solve for the variable by scaling or representing the problem on a number line.

(Possible solution strategies shown.)

⑨ $7 \cdot \frac{2}{7}m = 20 \cdot 7$

$2m = 140$

$m = 70$

⑩ $\frac{3}{5} + \frac{1}{10} = k$

$\frac{6}{10} + \frac{1}{10} = k$

$\frac{7}{10} = k$

Determine the missing value in each of these equations.

⑪ $\frac{22}{16} = \frac{a}{8}$ $a =$ 11

⑫ $\frac{35}{b} = \frac{49}{7}$ $b =$ 5

⑬ $\frac{c}{8} = \frac{7}{2}$ $c =$ 28

⑭ $\frac{9}{27} = \frac{5}{d}$ $d =$ 15

⑮

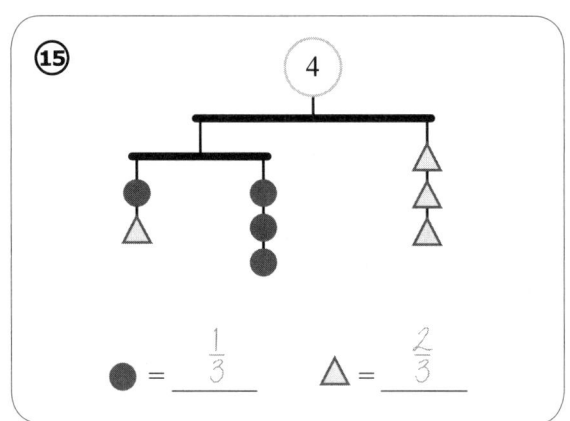

● = ___$\frac{1}{3}$___   △ = ___$\frac{2}{3}$___

⑯ Carla is making a bowl of punch for a birthday party. To make the drink, she mixes a 6-ounce can of frozen pink lemonade and 4 cups of lemon-lime soda. $p = 6$ $s = 4$

ⓐ Write an equation for the relationship between p, the number of ounces of pink lemonade, and s, the number of cups of soda.

$\frac{p}{s} = \frac{6}{4}$ or $\frac{s}{p} = \frac{4}{6}$ or $6s = 4p$ or $\frac{p}{s} = \frac{3}{2}$ etc.

ⓑ The store Carla visits only sells frozen pink lemonade in a 9-ounce can. How many cups of soda should she use?

$\frac{p}{s} = \frac{6}{4}$

$\frac{9}{s} = \frac{6}{4}$

6 cups

ⓒ Carla uses the same recipe to make punch for a school-wide event. She uses a giant can of frozen pink lemonade that contains 20 ounces. How many cups of soda should she use?

$\frac{p}{s} = \frac{6}{4}$

$\frac{20}{s} = \frac{6}{4}$

$\frac{80}{6}$ cups

(that's also $\frac{40}{3}$ cups)

Unit 8

Mental Mathematics

Fractions, approximation, and factors

To classify fractions as less than, equal to, or greater than $\frac{1}{2}$ (or $\frac{1}{4}$), one must see whether their numerator is less than, equal to, or greater than half (or one-fourth) of the denominator. The arithmetic task is a rehearsal of earlier learning, but this context builds a new sense of fractions. Assessing how many halves or thirds are in various numbers builds the sense that dividing by a fraction is like multiplying by its reciprocal. Students begin to make division approximations as well. First, they use a power-of-10 approximation to consider how many 23's fit inside 1000, and then, continuing an earlier theme, they refine those approximations.

Students also become familiar and make associations with factor pairs and common factors.

In nearly all of these mental mathematics activities, students "enact a function": an input-output rule is established at the outset, and students give the output for each input they hear. Each function rule focuses on a key mathematical idea or property (e.g. complements or the distributive property) that students begin to feel intuitively.

After introducing the day's task, the teacher deliberately does not reiterate the task but says only the input numbers for students to transform. Minimizing words lets students focus on the numerical pattern of the activity, helping them perceive the structure behind the mathematics. A lively pace maximizes practice and keeps students engaged.

Mental Mathematics • Activity 1
Comparing fractions to $\frac{1}{2}$

PURPOSE

Students classify positive fractions as $x < \frac{1}{2}$, $x = \frac{1}{2}$, or $x > \frac{1}{2}$, with special attention to the boundary case $x = \frac{1}{2}$. In doing so, they develop a sense of the magnitude of fractions and strengthen their understanding of the relationship (and distinction) between numerator and denominator. All of their skill in doubling and halving now has a new use: detecting fractions that are equivalent to $\frac{1}{2}$ and recognizing ones that are greater or less than $\frac{1}{2}$.

Some useful fractions to classify include $\frac{5}{8}$, $\frac{3}{8}$, $\frac{7}{16}$, $\frac{9}{16}$, $\frac{7}{12}$, and $\frac{5}{12}$, because they help students consider cases close to $\frac{1}{2}$.

Instructions:

Display (or draw on the board) the table shown on the following page. Ask students up, two or three at a time, to think of two fractions that belong in different categories and write them in the correct place. They must not duplicate a fraction that is already written (though they may write a different fraction having the same value). Be sure to explain the notation of the different columns (e.g. $0 < x < \frac{1}{2}$ means the fraction should be between 0 and $\frac{1}{2}$). When all students have had a chance to write something, work together as a class to find a few more good entries for the emptier column(s). Use the denominators for the middle column ($x = \frac{1}{2}$) to help students place fractions in the other columns. For example, if $\frac{4}{8}$ is placed in the middle column, ask students to place $\frac{3}{8}$ and $\frac{5}{8}$. Refer them to the $x = \frac{1}{2}$ column for comparison.

About this sorting activity:

By choosing a numerator and doubling or choosing a denominator and halving, one creates a fraction that is equal to $\frac{1}{2}$. Changing either the numerator or denominator changes their relationship. If the numerator is *more* than half the denominator, the fraction is *greater* than $\frac{1}{2}$. And if the numerator is *less* than half the denominator, the fraction is *less* than $\frac{1}{2}$.

An example of a completed activity is shown below:

$0 < x < \frac{1}{2}$	$x = \frac{1}{2}$	$\frac{1}{2} < x < 1$
$\frac{3}{7}$ $\frac{3}{8}$ $\frac{1}{4}$ $\frac{2}{6}$ $\frac{46}{100}$ $\frac{1}{3}$ $\frac{4}{10}$ $\frac{1}{8}$	$\frac{3}{6}$ $\frac{50}{100}$ $\frac{2}{4}$ $\frac{4}{8}$ $\frac{5}{10}$ $\frac{3\frac{1}{2}}{7}$	$\frac{6}{10}$ $\frac{5}{9}$ $\frac{5}{7}$ $\frac{5}{8}$ $\frac{3}{4}$ $\frac{2}{3}$ $\frac{73}{100}$ $\frac{7}{8}$

Comparing fractions to $\frac{1}{2}$ (Mental Mathematics Activity 1)

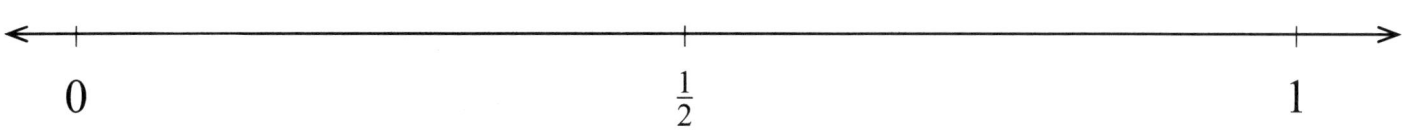

$0 < x < \frac{1}{2}$	$x = \frac{1}{2}$	$\frac{1}{2} < x < 1$

Mental Mathematics • Activity 2
Comparing fractions to $\frac{1}{4}$

PURPOSE
Students classify positive fractions that are all less than one-half as $x < \frac{1}{4}$, $x = \frac{1}{4}$, or $x > \frac{1}{4}$, with special attention to the boundary case $x = \frac{1}{4}$, further refining their understanding of the magnitude of fractions and how that is reflected in the relationship between the numerator and denominator. Again, they apply their skill in doubling and halving in a new context.

Instructions

Display the following page and allow students to come up with their own fractions to categorize. To encourage variety, ask students not to duplicate each other's responses. When all students have had a chance to write something, work together as a class to fill out the emptier column(s). Use the denominators for the middle column ($x = \frac{1}{4}$) to help students place fractions in the other columns. For example, if $\frac{2}{8}$ is placed as a value equivalent to $\frac{1}{4}$, ask students to place $\frac{1}{8}$ or $\frac{3}{8}$ in the proper categories. Refer to the $x = \frac{1}{4}$ column for comparison.

About this sorting activity:

By choosing a numerator and multiplying by 4 (say, by doubling twice), or choosing a denominator and dividing by 4, one creates a fraction that is equal to $\frac{1}{4}$. Changing either the numerator or denominator changes their relationship. If the numerator is *more* than one-fourth of the denominator, the fraction is *greater* than $\frac{1}{4}$. And if the numerator is *less* than one-fourth of the denominator, the fraction is *less* than $\frac{1}{4}$.

An example of a completed activity is shown below:

$0 < x < \frac{1}{4}$		$x = \frac{1}{4}$		$\frac{1}{4} < x < \frac{1}{2}$	
			$\frac{1}{4}$		
$\frac{1}{15}$	$\frac{4}{20}$	$\frac{5}{20}$	$\frac{2}{8}$	$\frac{3}{8}$	$\frac{7}{17}$
$\frac{1}{8}$	$\frac{3}{16}$	$\frac{4}{16}$			$\frac{6}{16}$ $\frac{7}{16}$
$\frac{3}{24}$					
$\frac{1}{12}$	$\frac{22}{100}$	$\frac{6}{24}$ $\frac{3}{12}$	$\frac{25}{100}$	$\frac{35}{100}$	$\frac{5}{12}$
			$\frac{7}{28}$	$\frac{8}{28}$	$\frac{9}{20}$

- Some useful fractions to work on as a class include $\frac{3}{8}$, $\frac{1}{8}$, $\frac{3}{16}$, $\frac{5}{16}$, $\frac{2}{12}$, and $\frac{5}{12}$, because these cases are near the boundary case.

- Thinking about the relationship between numerator and denominator helps us classify fractions like $\frac{7}{17}$. The numerator is less than half the denominator, so $\frac{7}{17} < \frac{1}{2}$, and it is more than $\frac{1}{4}$ of the denominator, so $\frac{7}{17} > \frac{1}{4}$. Students might also think, "I know $\frac{7}{16} < \frac{1}{2}$ because $\frac{7}{16} < \frac{8}{16}$, and I know that $\frac{8}{16}$ is *equal to* $\frac{1}{2}$. Seventeenths are smaller than sixteenths, so $\frac{7}{17} < \frac{7}{16} < \frac{1}{2}$. And is it less than a fourth? It can't be, because $\frac{7}{28}$ equals $\frac{1}{4}$, and twenty-eighths are smaller than seventeenths, so $\frac{7}{28} < \frac{7}{17} \ldots$" and so on. Both kinds of reasoning are excellent.

Comparing fractions to $\frac{1}{4}$ (Mental Mathematics Activity 2)

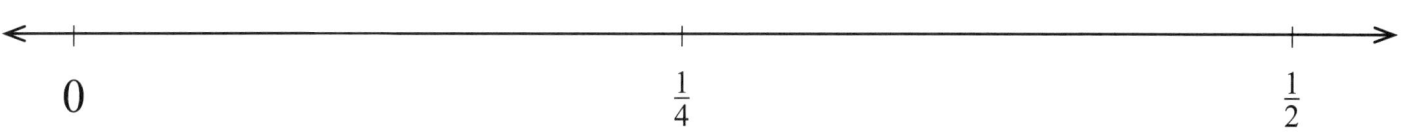

$0 < x < \frac{1}{4}$	$x = \frac{1}{4}$	$\frac{1}{4} < x < \frac{1}{2}$

How many halves?

13   26

Mental Mathematics • Activity 3
How many halves? How many thirds?

PURPOSE

One interpretation of $c \div b$ is "How many b in c?" This activity provides one logical foundation for dividing by fractions. Let students discover, on their own, that dividing a number by $\frac{1}{2}$ gives the same result as doubling that number.

Introduce:

"Whatever number I say, you say how many halves are in the number. How many halves in 1? Of course, there are 2. How many halves in $1\frac{1}{2}$? Yes, 3. What if I say 4? How many halves is that? Yes, 8. Got it? Here we go."

- Make the connection that dividing by $\frac{1}{2}$ answers "How many halves?" It may help to first remind students of the correlation between asking "How many … are in …?" and division by discussing a similar question with integers, e.g. considering how to answer the question "How many sevens are in 28?" which is answered by dividing by 7. Let students notice that dividing by $\frac{1}{2}$ is the same as multiplying by 2.
- Similarly, dividing by $\frac{1}{3}$ answers "How many thirds?" If there is time, let students discuss why.
- Examples like $\frac{12}{2} \div \frac{1}{2} = 12$ and $\frac{6}{3} \div \frac{1}{3} = 6$ connect closely with $\frac{12}{2} \cdot 2 = 12$ and $\frac{6}{3} \cdot 3 = 6$, which are important ideas in the unit. For added challenge, you might discuss $4 \div \frac{1}{3} = 12$ and $4 \div \frac{2}{3} = 6$.

Step 1: Ask, "How many halves?" Include integers and mixed numbers with halves.

3	6	$\frac{1}{2}$	1
5	10	**11**	22
$5\frac{1}{2}$	11	**$11\frac{1}{2}$**	23
$1\frac{1}{2}$	3	**13**	26
$2\frac{1}{2}$	5	**$6\frac{1}{2}$**	13
$3\frac{1}{2}$	7	**$12\frac{1}{2}$**	25
4	8	**15**	30
$4\frac{1}{2}$	9	**16**	32
2	4	**$14\frac{1}{2}$**	29
1	2	**$8\frac{1}{2}$**	17
$\frac{3}{2}$	3	**$\frac{12}{2}$**	12
$\frac{5}{2}$	5	**6**	12

Step 2: "Let's switch to thirds. How many thirds are in 1? What about 2? Okay, from now on, whatever number I say, you say how many thirds in it."

$\frac{1}{3}$	1	**0**	0
$\frac{2}{3}$	2	**5**	15
$1\frac{1}{3}$	4	**$5\frac{2}{3}$**	17
$2\frac{2}{3}$	8	**7**	21
3	9	**$6\frac{1}{3}$**	19
$3\frac{1}{3}$	10	**9**	27
10	30	**$9\frac{1}{3}$**	28
$10\frac{2}{3}$	32	**12**	36
33	99	**$12\frac{2}{3}$**	38
$33\frac{1}{3}$	100	**$\frac{10}{3}$**	10
11	33	**$3\frac{1}{3}$**	10
$\frac{2}{3}$	2	**$\frac{6}{3}$**	6
$\frac{12}{3}$	12	**2**	6

Extension: "Here's a challenge! You know how many *thirds* there are in 4, right? Yeah, 12 of them. So how many '*two-thirdses*,' pieces *twice* the size of one-third, are there in 4?" (6)

Mental Mathematics • Activity 4
Tripling

PURPOSE

In doubling and halving, students have broken seemingly complicated calculations into simpler steps. In this activity, students apply that skill to tripling 2-digit integers, using the distributive property again.

Introduce:

"Any number I say, you multiply by 3. So if I say 5, you say 15. Got it?"

Step 1: Have students multiply by 3. Use single-digit numbers.

1	3	**8**	24
4	12	**-2**	-6
6	18	**9**	27
2	6	**7**	21
8	24	**3**	9
5	15	**6**	18
-4	-12	**9**	27
7	21	**4**	12
-3	-9	**2**	6

Step 2: Include 2-digit and simple 3-digit numbers scaffolded by simpler inputs like multiples of 10 and single-digit numbers.

6	18	**20**	60
60	180	**7**	21
63	189	**27**	81
60	180	**28**	84
5	15	**26**	78
65	195	**25**	75
-65	-195	**40**	120
600	1800	**41**	123
602	1806	**47**	141

Mental Mathematics • Activity 5
Approximation

PURPOSE
The first step in division is an approximation of how many of one number "fit in" another. This activity builds skill at that first step. It also supports students' skill at multiplying by 10 and 100.

Introduce:

"Roughly, *just approximately*, how much is 1000 ÷ 23? That is, how many twenty-threes are in 1000? Certainly more than 1! Are there at least 10? Sure. Ten of them are only 230. Can we fit 100? No? So, 1000 ÷ 23 is somewhere between 10 and 100. Later, you will learn how to refine your mental approximations, but for now, between 10 and 100 is good enough. Ready?"

- If students excel at this, they're ready to refine their approximations.
- If not, stop for today, but repeat this activity with new numbers another time. Introduce the repeat activity with a re-explanation of the task. "How do we figure out how many nineteens are in 5774? Well, *10* nineteens is 190, so *10* is too small. *100* nineteens is 1900; that's still less than 5774. *1000* nineteens is 19,000. That's *more* than 5774, so 5774 ÷ 19 is somewhere between 100 and 1000."

Write the division on the board and then ask, "How many _____ are in _____?" followed immediately by "Can you fit at least 1?" Allow students time to respond, and then ask "At least 10?" and then "100?" If the answer is "yes" to all of those, ask, "1000?" or, in the final case, "10,000?" Each time, summarize for students the boundaries of their estimate. For example, after they have responded "no" to 3000 ÷ 17, have them summarize with "So, 3000 ÷ 17 is somewhere between 100 and 1000."

		At least 1?	10?	100?	1000?
200 ÷ 3	How many threes in 200?	yes	yes	no	
300 ÷ 17	How many seventeens in 300?	yes	yes	no	
3000 ÷ 17	How many seventeens in 3000?	yes	yes	yes	no
2000 ÷ 9	How many nines in 2000?	yes	yes	yes	no
2000 ÷ 4	How many fours in 2000?	yes	yes	yes	no
400 ÷ 27	How many twenty-sevens in 400?	yes	yes	no	
2000 ÷ 76	How many seventy-sixes in 2000?	yes	yes	no	
3000 ÷ 132	How many one hundred thirty-twos in 3000?	yes	yes	no	
5774 ÷ 19	How many nineteens in 5774?	yes	yes	yes	no
2014 ÷ 2	How many twos in 2014?	yes	yes	yes	yes

Mental Mathematics • Activity 6
Closer approximation

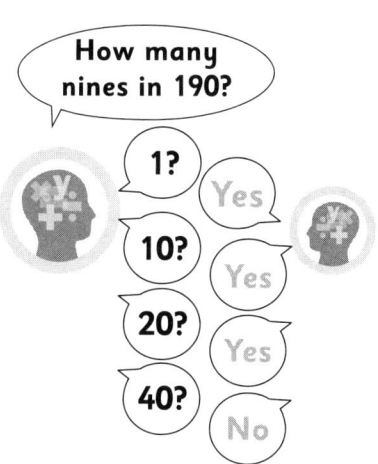

PURPOSE
Students learn that it is possible to refine approximations. They also see how the mental shortcuts they have learned help them reason about familiar mathematics (in this case, division). In this activity, students apply their skills at multiplying by 10 and doubling to make closer approximations in division.

Introduce:

"Mental division again today—we'll still just approximate, but we'll take one more step. For example, to do 237 ÷ 6, we ask how many sixes are in 237? More than 1. What about 10? Sure. 100? No. So, it's between 10 and 100. How much is 10 sixes? (60) What about 20 sixes? How much is that? (120) Still too little. What about 40 sixes? (240) *Just barely* too big. So 237 ÷ 6 is somewhere between 20 and 40, but it's really close to 40.

"Yesterday, we would have stopped somewhere between 10 and 100. Today we got closer, and without doing a lot of hard work. And we can keep getting more precise, until we are as close as we want. Ready for more?"

About this sequence:

For each division problem in this activity, begin with the "powers of 10" prompts ("1?" "10?" "100?" etc.), and stop those once the answer is "no." Then, articulate or have a student articulate that "The answer is between _____ and _____ " and refine by prompting with either "2?" "4?" and "8?" or "20?" "40?" and "80?" Before moving on to the next problem, have someone describe the refined approximation: "The answer is between _____ and _____. "

- Remind students that they are approximating division by using multiplication. That is, to see if there are at least 20 elevens in 300, they multiply 11 by 20 and consider that 220 is less than 300 so there must be more than 20 elevens.
- Revisit doubling and multiplying by 4, 10, 20, 40, and/or 100 if your students struggle with this activity.
- If the answer to "Are there at least 40?" is "yes," you may choose to ask, "Can you fit at least 80?" or settle on "between 40 and 100" as the approximation.

		1?	2?	4?	10?	20?	40?	100?
300 ÷ 11	How many elevens in 300?	yes			yes	yes	no	no
297 ÷ 7	How many sevens in 297?	yes			yes	yes	yes	no
310 ÷ 4	How many fours in 310?	yes			yes	yes	yes	no
599 ÷ 143	How many one hundred forty-threes in 599?	yes	yes	yes	no			
800 ÷ 242	How many two hundred forty-twos in 800?	yes	yes	no	no			
4000 ÷ 231	How many two hundred thirty-ones in 4000?	yes			yes	no		no
800 ÷ 33	How many thirty-threes in 800?	yes			yes	yes	no	no
800 ÷ 13	How many thirteens in 800?	yes			yes	yes	yes	no

Factors of 30

Mental Mathematics • Activity 7
Factor pairs

PURPOSE
Students have focused a lot on decomposing numbers into *sums* of other numbers. By rehearsing some common factor pairings, students begin to develop the habit of looking for multiplicative decompositions. That is, they learn to see numbers as *products* of other numbers.

Introduce:

Write 20 on the board. "We can make this integer, 20, by adding various pairs of numbers. For example, $18\frac{1}{2} + 1\frac{1}{2}$. We can also make it by *multiplying* two numbers like 20 • 1. Because those are *both integers* whose product is 20, we call them *factors* of 20.

"The number 20 can be factored in other ways. I'll say a number. If there is an integer that you can multiply that number by to get 20, you name it. So, if I said 4, you'd say... 5, of course. What if I said 8? You *can* multiply 8 times a number to make 20, because $8 \cdot 2\frac{1}{2} = 20$, but that number isn't an integer, so 8 is not a factor of 20. So, if I give you a number that is not a factor, then just say 'No.'

"Let's try making factor pairs to 20. If I said 2, you'd say... 10. If I said 3, you'd say... 'No.' I'll keep saying numbers. When we have found all the factor pairs, you can say 'Done!'" Write the factor pairs on the board as they are found.

- When students decide that all factors have been found, whether or not they are correct, give them a bit of time to figure out if they are certain. This can liven up the activity and help students come up with their own process for deciding when all factors have been identified. It is also playful, as these activities should be.
- You may wish to ask students to share strategies for knowing when all factors have been identified.
- When checking integers sequentially, it is sufficient to check all numbers up to and including half of the product in question.

Factors of 12	
1	12
2	6
3	4
4	3
5	no
6	2
7	no
8	no

Factors of 18	
1	18
2	9
3	6
4	no
5	no
6	3
7	no
8	no
9	2
10	no

Factors of 36	
1	36
2	18
3	12
4	9
5	no
6	6
7	no
8	no
9	4
10	no

Factors of 28	
1	28
2	14
3	no
4	7
5	no
6	no
7	4
8	no
9	no
10	no

Extension: Factors of 75			
1	75	**4**	no
2	no	**5**	15
3	25	**6**	no

Mental Mathematics • Activity 8
Finding all factors

PURPOSE
Students explore the process of finding all factors of a number. The purpose is to nudge students toward a systematic way of knowing when they have found all the factors.

Introduce:

"Yesterday, we found factor pairs. Today, whatever number I write, your job is to list *all* of its factors. I'll write the factors as you say them, and when you believe that there aren't any more to find, say 'Done.' For example, if I say 8, you say… (1, 2, 4, 8). Any other factors? Done? Ready?"

About this sequence:

The factors are listed here in order, but of course students can identify them in any order and may list each factor with its pair.

Step 1: Write the input on the board, and record the factors as students identify them.

24	1	2	3	4	6	8	12	24
16	1	2	4	8	16			
36	1	2	3	4	6	9	12	18	36
45	1	3	5	9	15	45		
32	1	2	4	8	16	32		

Extension: If there is time, consider two powers of 10.

10	1	2	5	10					
100	1	2	4	5	10	20	25	50	100

- When students say "Done," ask them to take a moment to consider if there are any other factors. You may wish to prompt for missing factors with questions such as "Did we list both numbers in each factor pair?" or "What are the factors of 6? Are those numbers factors of 36?" It can be interesting to discuss that the matching factor for 4 when considering factors of 16 is itself. Students may benefit from considering whether the factors of a factor are always factors of the product in question (e.g. if 6 is a factor, does that mean 2 and 3 always are?). You may also prompt with numbers that are not factors, asking questions like "Is 5 a factor of 16? How do you know?"
- Students may miss the factor 1 and/or the product itself. This may be a misunderstanding about factors or just an oversight. Have students describe the definition of a factor in their own words and consider these numbers with that definition in mind.

Common factors of 14 and:

Mental Mathematics • Activity 9
Common factors

PURPOSE

As students develop associations between numbers and their factors, they become able to find commonalities. This activity builds students' skill at looking for common factors, which is often used in algebra.

For each pair of numbers, write "Common factors of... and..." on the board. With each new prompt, change the numbers as necessary.

Introduce:

"I'll give you two numbers, and you name any factors they have in common. So if I say 12 and 18, you say... yes, 1, 2, 3, 6.... Any others? Okay, let's try some different numbers."

About this sequence:

Students first consider common factors between 24 and another number and then between 15 and another number. In both steps, students must mentally juggle the factors of two separate numbers, hold them in mind long enough to compare them, and call out only the ones that match. The oral format of the activity is more challenging than finding common factors on paper, but this format allows students to focus on the associations they have made between numbers and their factors, and builds working memory in ways that a pencil-and-paper format does not.

Step 1: Ask students to identify the common factors of **24** and:

18	1	2	3	6		
20	1	2	4			
15	1	3				
12	1	2	3	4	6	12
10	1	2				

Extension: Common factors of **21** and:

15	1	3
36	1	3
49	1	7

Step 2: Ask students to identify the common factors of **15** and:

30	1	3	5	15
40	1	5		
18	1	3		
21	1	3		
45	1	3	5	15
10	1	5		
9	1	3		
20	1	5		

Mental Mathematics • Activity 10
Common factors (day 2)

Common factors of 36 and:

15   3

1   done

PURPOSE

This activity continues to focus on developing the useful mathematical skill of looking for and recognizing common factors.

Introduce:

"Common factors again today! So if I say 45 and 15, you say… yes, 1, 3, 5, and… 15. Are we done? Are you sure? How do you know? Okay, let's try more."

Again, write each pair of numbers on the board so students can focus on finding common factors instead of remembering what you just said.

Step 1: Ask students to identify the common factors of **36** and:

20	1	2	4
15	1	3	
21	1	3	
45	1	3	9
56	1	2	4
49	1		

Step 2: Ask students to identify the common factors of **48** and:

10	1	2		
14	1	2		
42	1	2	3	6
35	1			
54	1	2	3	6
100	1	2	4	

Extension: Ask students to identify the common factors of sets of three numbers.

6, 18, and 36	1	2	3	6
10, 100, and 1000	1	2	5	10

Mental Mathematics • Activity 11
Choose a review

PURPOSE

At this point in the unit, students have learned a lot of new things and worked with several different Mental Mathematics formats. Take some time to review a particularly challenging, fun, or fruitful activity and allow students to achieve and enjoy mastery. Some suggestions are listed below.

About this selection:

All of the activities in this unit (and any other, for that matter) are worthy of review. The three listed below particularly embody some of the mental habits for students to internalize.

Comparing fractions by comparing to known quantities (like $\frac{1}{2}$ and $\frac{1}{4}$) builds comfort with fractions and a deeper understanding of the information contained in the numerator versus the denominator. Calculating thirds strengthens one's sense of the distributive property (beyond doubling and halving) and builds toward an understanding of division by fractions. The approximation activity makes clear to students how they can judiciously substitute easy calculations for difficult-seeming ones.

Even though the activities relating to factoring help students to think about decomposing numbers as products and they assist students in recognizing important common factors, they are less suited as review choices for two reasons. It is harder to keep these fresh and challenging during review, and their focus is not on key algebraic properties as much as it is on particular arithmetic knowledge.

Comparing fractions to $\frac{1}{4}$: Revisit Activity 2. You may reuse the inputs from that activity or come up with new ones. Remember not to introduce too many new or challenging inputs, as this is intended to be review and a chance for students to experience mastery.

How many thirds: Rerun Activity 3, Step 2, and have students tell you how many thirds are in each number. You may wish to ask students to make explicit the connection between this and Activity 4: Tripling

Closer approximation: Revisit Activity 6. Use similar inputs to the ones listed and allow students to practice mental division and approximation. Again, the object is for students to feel their own mastery of this useful skill and way of thinking, so give challenge but not beyond students' comfort zone.